Herbert Lang

Matte Smelting

Its principles and later developments discussed, with an account of the

pyritic processes. Second Edition

Herbert Lang

Matte Smelting
Its principles and later developments discussed, with an account of the pyritic processes. Second Edition

ISBN/EAN: 9783337312305

Printed in Europe, USA, Canada, Australia, Japan

Cover: Foto ©berggeist007 / pixelio.de

More available books at **www.hansebooks.com**

MATTE SMELTING.

ITS PRINCIPLES

AND

LATER DEVELOPMENTS DISCUSSED.

WITH AN ACCOUNT OF THE

PYRITIC PROCESSES.

BY

HERBERT LANG

SECOND EDITION.

NEW YORK AND LONDON:
THE SCIENTIFIC PUBLISHING CO.,
1898.

INTRODUCTION.

IN ITS present development matte-smelting is applied in the extraction of gold, silver, copper, nickel, cobalt, and lead from their ores. It is probable that more than one-half of the world's supply of copper is obtained in this way, while the proportion of silver thus procured is very large and is yearly increasing. We do not possess precise statistics bearing on the subject, but approximate estimates appear to show that the aggregate value of the metals which are extracted annually throughout the United States by means of matte-smelting methods has now reached the magnificent total of thirty millions of dollars. The interest that is naturally felt in metallurgical processes which are accomplishing such vast results is increased by the fact that they are in a condition of rapid improvement and expansion, exhibiting at the present moment a vitality and a progressiveness as great, perhaps, as is elsewhere shown in the whole range of metallurgy. We may confidently expect not only a higher perfection in their application to the metals of the foregoing list, but also the extension of the principles of the art of matting to the benefication of ores of other metals and metalloids. It is not unreasonable to expect that in the perhaps immediate future we may by such means recover arsenic, antimony, tin, bismuth, and the metals of the platinum group; and that by modifications and combinations of already known processes, sulphur itself may be practically recovered as it issues from the flues of the matting furnace. Most metallurgists will doubtless coincide in the assertion that matte-smelting is therefore unequaled in the variety and extent of its applications, as well as in its probable future expansion, by any other process, or system of processes, known to their art.

Matte-smelting, in common with other departments of ore

reduction, has undergone changes and improvements of great
magnitude within recent years—changes and improvements which
are not fully recorded in our text-books, and of which the
ordinary reader and the merely book-learned individual can have
but slight conception. The smelting of to-day is far different
from the smelting of twenty, of ten, or even of five years ago, and
even those metallurgists who, like the writer, have been actively
employed in such pursuits are themselves as yet hardly able to
realize the full bearing and application of the principles and processes so lately worked out.

It may be that a time of transition like the present, when discovery heralds discovery, and improvement merges into improvement, is not the best time for writing the finished treatise which
would mark the closing of an era in metallurgy. But with less
ambition, less leisure, and perhaps less fitness for the task, still
one may doubtless find a great deal to say which would be profitable to write and to read. There are many topics connected with
smelting which, being unfitted by their purely practical nature
from discussion in books, we have had no formal instruction upon.
Others there are which have escaped discussion because of their
novelty. It will be the effort in the following essay to touch upon
topics of this practical sort, and to bring those of an isolated or
novel kind into correlation with the underlying principles of
metallurgy, as it is certain that no discussion worthy of the name
can ignore the theoretical basis of the subject. While endeavor-
ing, therefore, to supply the want of present information in particular lines of work and research, the writer would disclaim any
attempt to produce a complete treatise, or one covering fully the
domain of matte smelting, certain important divisions of which,
as, for example, reverberatory furnace smelting, remaining un-
touched in this writing.

A question having arisen during a year or two past as to the
uses and comparative efficiencies of the matting and the lead
smelting processes, and the query having especial pertinence to
this subject, I think it well to place this discussion before the
reader somewhat in the form of a comparison of the practice,
and incidentally of the principles of the two related arts. As a
basis of an understanding of their salient differences it is convenient to classify them as: 1. Difference in the carrier; 2. Difference
in the slags; 3. Difference in the fuels; 4. Difference in apparatus;

5. Difference in materials treated. Accordingly I shall adopt such a classification of topics in the ensuing essay.

TREE SHOWING RELATIONS OF THE MATTING PROCESSES.

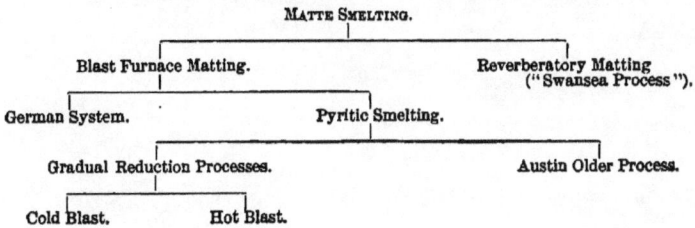

MATTE SMELTING.

Blast Furnace Matting.   Reverberatory Matting
                                                ("Swansea Process").

German System.   Pyritic Smelting.

Gradual Reduction Processes.   Austin Older Process.

Cold Blast.   Hot Blast.

CONTENTS.

8   CONTENTS.

SECTION FOUR.—LOSSES IN SMELTING, SALE OF PRODUCTS,
TABLES.

MATTE SMELTING.

SECTION ONE.

THE CARRIER.

1. THE CARRIER CONSIDERED.—In lead smelting, as all are aware, the carrier is metallic lead, and the precious metals come down as alloys with lead, intermingled with the great excess of that metal. But inasmuch as other valuable metals which we may wish to save, as copper, nickel, and cobalt, do not form with lead the manageable alloys which would enable their simultaneous extraction, and as the conditions necessarily attending their reduction to the metallic form are incompatible with the saving of the lead, we are debarred from the employment of that method for their extraction. Although all the substances mentioned, as well as others, cannot be simultaneously produced as metals, their artificial sulphides may all be produced at one and the same operation, and the metals won may be separated from each other and from attending impurities by a series of processes analogous to and not more difficult or costly than those by which lead bullion is converted into marketable metals. This smelting operation, which is called by the Germans "Raw Smelting," and in this country is indifferently known as "Matting," "Matte Smelting," or "Pyritic Smelting," seems to merit better the name "Sulphide Smelting," as a more distinctive and logical designation. While preferring the latter term I shall in deference to established custom continue to use the term "Matte Smelting," as in the present writing.

2. DEFINITION OF MATTE SMELTING.—Matte smelting may be defined briefly as the smelting of natural sulphides with the design of collecting their valuable parts in a quantity of artificial sulphides. Or, more explicitly, as the smelting of ores composed

of, containing, or giving rise to, sulphides, for the purpose of collecting their values in a less quantity of artificial sulphides. And since, as will be shown, arsenic or antimony can take the place of sulphur in the ore mixture with entire success to the operation, we may and sometimes do have arsenide and antimonide smelting; for which the above definition, with the obvious changes, will answer perfectly. Matte, the name by which is designated the indefinite mixture of artificial sulphides which are the result of the smelting operation, covers many varieties of product, varying infinitely in chemical composition and physical characteristics. One of the most striking features, and one with which the smelter is very deeply concerned, is the miscibility of mattes; whereby we find sulphides of even considerable differences in specific gravity, and in fusibility, and of immense difference in chemical composition, mixing together in the liquid state with such thoroughness that no characteristic peculiar to, or even suggestive of, either can be recognized in the mingled mass. It may be that this excessive miscibility partakes of the nature of solution, one sulphide being virtually dissolved in another, as sugar dissolves in water. As to this point I shall present a few remarks in another connection.

3. How Mattes are Classified.—No classification of these very remarkable compounds has been made other than their designation according to the predominant metal—or oftentimes according to the metal of predominant value—into iron matte, copper matte, lead matte, silver matte, etc. It will assist the effort to obtain a logical and scientific view of the subject if we endeavor to classify them in a manner more consistent with the present condition of metallurgical science. I would suggest that there is nothing in the composition of any of the mattes, or even in the composition of the arsenide and antimonide compounds which we know by the objectionable term speisses, to prevent their being brought into a family together, and considered as individual members of a great class. The classification under which I prefer for the present purpose to place both the mattes and the speisses is as sulphide mattes, arsenide mattes, and antimonide mattes. Examples of each will be found in their appropriate places in the Table of Smelting Products accompanying this article. This classification I shall continue to follow, dropping the term speiss as distinctive of nothing which we could wish to preserve in our speech. I shall use the less limited term matte in the general

sense of meaning either the sulphide, the arsenide, or the
antimonide compound, or the mixtures of all of them, according to
circumstance.

4. THE COMPOSITION OF MATTES.—Of the common metals
which play an important part in the chemical composition of mattes,
there are iron, copper, lead, zinc, nickel, and cobalt; while of the
less common ones, the most important are silver, gold, and bismuth.
Occasionally manganese and tin occur; and even the comparatively rare elements, vanadium, molybdenum, cadmium, and
platinum have been detected. Even the metals of the alkaline
earths enter at times into the composition of mattes, and, as will
be shown, they sometimes play not unimportant parts therein.
Those chemical analyses of mattes to which I have access demonstrate profound differences of composition, which I can best illustrate by the following citations. Of the various elements which
enter into the composition of certain mattes, I quote the highest
percentages and the lowest which are found therein:

	Highest.	Lowest.		Highest.	Lowest.
Iron	70.47	0.136	Platinum	0.0018	0.
Copper	80.	0.	Bismuth	1.26	0.
Lead	73.	0.	Molybdenum	2.31	0.
Zinc	11.5	0.	Calcium	7.	0.
Nickel	55.	0.	Barium	22.	0.
Cobalt	54.	0.	Sulphur	44.	trace
Manganese	8.	0.	Arsenic	52.	0.
Silver	5.	0.	Antimony	60.	0.
Gold	0.11	0.			

An inspection of this table renders obvious the prodigious
diversity of composition of mattes; whence we may infer the application of the matting processes to widely differing ores and
mixtures. We generalize also that no single element of the list
is indispensable in matte formation, but that the place of each
may be taken by others. Even sulphur and iron, which are the
most common and abundant constituents, and which appear in
every matte whose analysis I have encountered, with a single
exception, are in many cases to be considered merely as accidental
impurities, and not as essential to the constitution.

5. VIEWS AS TO CHEMICAL COMPOSITION.—Balling and
others have derived from various analyses the formulæ Fe_2S_3,
FeS, Fe_2S, and even Fe_4S, as representing the various states
in which iron exists in sulphide mattes; while the average of

opinion and experience favors FeS as the usual combination. Possibly all of these combinations may have been found; but uncertainty exists which ill accords with the advanced condition of chemical science.

But if our knowledge as to the mode of combination of the iron in the sulphide mattes is uncertain and unsatisfactory, it is still more so in reference to the arsenide compounds of that metal. Observers have recognized, or claimed to recognize, a long series of ferrous arsenides of artificial origin, not less than a dozen in number, differing from each other in well-marked physical as well as chemical characteristics, and all producible at will by the instrumentalities of the blast furnace. Arsenide mattes are especially prone to assume crystalline forms on cooling, and it is peculiarly pertinent to remark here that a tap of that material freqently separates into two parts, one thoroughly crystallized at the bottom and with a massive or incipiently crystallized portion above.

Analysts are unanimous in ascribing the familiar composition Cu_2S to the copper compound which we find in sulphide matte, and an equivalent ratio to the same metal in its combinations with arsenic and antimony. In assuming the unvarying character of its ratio of combination it appears that we stand upon firm ground, and a safe starting place from which, if we had analytical evidence enough, we might proceed to the discussion of the general questions of matte formation. Certain other elements, as nickel, cobalt, and manganese, as far as is known, comport themselves with no less strictness than copper, forming no compounds other than those analogous in composition with their native mono-sulphides. As to lead; metallurgists have taken it for granted that this metal possesses but one sulphide, PbS; but the view of theoretical chemists favors the possibility of the existence of the two lower sulphides, Pb_2S and Pb_4S, in matte.

Of metallic substances in matte, we are all familiar with the existence of copper in the metallic form in rich furnace products of that kind. Guyard has isolated metallic iron and metallic lead from the mattes of Leadville; and Mr. Pearce has proved the existence of metallic gold in certain experimentally formed sulphides. I have myself noticed the volatilization of metallic zinc from molten matte, a phenomenon whose significance I am as yet unable to grasp.

Guyard's view that matte is essentially a combination of ferrous mono-sulphide (Fe S) with the magnetic oxide of iron, has not found many supporters; but that the latter substance always exists in matte there is reason for believing. He found 16 and 23 per cent. of magnetic oxide respectively in two samples of Leadville blast-furnace matte, using methods of separation which were somewhat unusual to chemists and concerning which we scarcely know enough to enable the proper weight to be attached to his conclusions. His assumption as to the composition of matte was weakened by his parallel one in reference to slag, which he defined as essentially a combination of silicates with calcium matte (*i.e.*, sulphide of calcium plus magnetic oxide of iron), which view likewise has failed of indorsement since its promulgation.*

6. THE GENERAL PROBLEM.—The general problem of the formation of matte in the furnace is an exceedingly important and interesting one, but is a problem, unhappily, that the present condition of metallurgy does not enable us to solve by any means as satisfactorily or discuss as luminously as could be wished. The problem naturally takes a form like this: Given a mixture of heavy metals and two metalloids, whose affinities at certain temperatures and under certain conditions are known. Required the resulting compounds when subjected to a much higher temperature. The problem, which, in its simplest form is totally incapable of exact solution in the light of chemistry, becomes vastly more complex when we realize the solvent power of the innumerable resulting compounds for unsaturated elements and for other compounds. In this problem we have a parallel for the well-known astronomical one of the Three Bodies, which transcends the keenest analysis of the mathematicians. Experiment long since taught us the respective affinities of the elements for oxygen; and later the researches of Fournet demonstrated the relative attractions of the heavy metals for sulphur; but the more difficult and recondite experimentation to establish the affinities of those metals toward both elements and toward each other at the smelting temperature has never been performed.

Chemical analysis, applied to the study of these interesting compounds, has not enabled us to fully understand their internal structure. Even regarding the simplest of them our analysis fails to substantiate our expectations. We find for example an

* *Mining Industry of Leadville,* Emmons, p. 724.

excess of iron in certain ferrous mattes, something beyond the amount which the doctrine of definite chemical proportions would teach us to expect; while the heterogeneous character of ferrous arsenides increases the difficulty of chemical analysis while diminishing the reliability of its conclusions. Analysis almost always shows unexpected excesses or deficiences of some elements not always to be ascribed to the effects of replacement by related elements. But this fact, be it noticed, does not destroy the analogy of these artificial compounds to native minerals, for these also rarely show by analysis the precise proportions demanded by theory.

7. THEORETICAL CONCEPTS.—In attempting to devise a theory of matte formation it is the most convenient and perhaps the most rational to conceive of all mattes as derived from the cuprous sulphide matte by replacement of copper successively by other metals whereby we obtain all the varieties of sulphide mattes; while the arsenide and antimonide mattes arise from the replacement of the sulphur by the two corresponding elements, arsenic and antimony. At first thought it would seem to the chemist that the ferrous sulphide matte is more truly typical of the class; and certain writers have even of late discussed mattes under the assumption that they are primarily compounds of iron and sulphur in which the former is replaced in part by lead and copper, and in less degree by zinc, silver, nickel, cobalt, manganese, arsenic, antimony, calcium, barium, and magnesium. In view, as before remarked, of the variable nature of the combination which iron forms with sulphur, and the uncertainty and lack of knowledge of their composition, it appears to me that the ferrous substances cannot conveniently or logically be taken as the archetype of mattes, but that the cuprous mattes should be so taken, if we assume any at all. Of all substances which enter matte, we must consider iron as one of the most uncertain in its combinations, and in so far ill calculated to serve as a starting point.

I prefer to look upon matte as a mixture of saturated compounds, each one of them equally important to its existence, and none of them indispensable to it. Matte would remain matte if any of its constituents were removed. Accordingly, I do not look upon the saving, for example, of gold and silver as the result of those metals becoming in some fortuitous manner entangled in matte, but rather as the result of the formation of double or multiple salts, into which the precious metals enter in definite proportion. However, beyond the force of chemical affinity which tends to definite combination, we are compelled to recognize, as shown by certain observed phenomena, which I need not here recount, the tendency to solution, which gives rise to indefinite mixture; and also the attraction of alloyage, or the propensity of certain reguline metals to form alloys. The subject is of an undoubtedly difficult and complex nature, such as the materials and facts at command in nowise enable us to treat adequately. I may suggest, in this connection, that as chemical analysis has not given us an adequate comprehension of the matter at issue, synthesis may do better. By proceeding synthetically in the direction of Dr. Pearce's experiments, previously cited, or upon a more ambitious working scale even, if such be found practicable, one can hardly fail to enlarge the bounds of metallurgical knowledge, while paving the way for practical results of the highest value. That experimenter is to be envied whose tastes and opportunities impel him upon this course of investigation.

It would be interesting indeed, were our knowledge sufficient, to trace the progress of a metal through the multifarious forms which, impelled by the play of chemical attraction, it assumes, as the furnace operation proceeds, from the moment when, as crude ore, it entered upon its baptism of fire until it emerges borne in fiery matte, or even when freed from " humors and corruptions" it attains the metallic form and is ready for the service of man; but for the present one can do the reader a better service by frankly declaring that from a reasonable point of view the results of such speculation, based as it must be upon our too limited knowledge of the inner harmonies of the smelting furnace, possess no practical value in comparison with the findings of experience.

8. ARSENIC AND ANTIMONY AS MATTE FORMERS. — The function of arsenic and antimony appears to be misunderstood. It is difficult to believe that these substances, which resemble sulphur in their modes of combination, can replace the heavy metals in mattes. We are familiar with both the native and the artificial sulphides of the two elements, and we find them reported in analyses as existing as such in mattes; but what we know of their physical characters, such as specific gravity and volatility, debars us from accepting the conclusion that they really are found therein. As an invariable rule, the specific gravity increases with the content of arsenic and antimony, while, should their sulphides exist, weighing as they do less than any

mattes, their effect upon the specific gravity would be the contrary of what is found. Nothing has been proved with more certainty than that the three elements, arsenic, antimony, and sulphur, may replace each other entirely in all the combinations which we meet in practical smelting. It appears most likely that arsenic in sulphurous mattes does not exist simply as sulphide, but as the sulpharsenide, the thio-arsenite of the chemists, formed by the combination of As_2S_3, with a sulphide of a heavy metal (e. g., PbS, As_2S_3). The type formula accordingly is R_2S, As_2S_3, in which R represents a monatomic metal. Under the same circumstances antimony acts similarly, forming sulphantimonides analogous to the first-mentioned substances, both series being chemically identical with their mineralogical relatives.

In ordinary estimation the arsenical and antimonial furnace products are held as follows:

1. Difficult of treatment by established processes. Especially is their calcination difficult, owing to the formation of oxidized non-volatile substances (arseniates and antimoniates).

2. Fused they corrode masonry; while those containing lead corrode iron through replacement of metals (e.g., $PbAs_n + Fe = FeAs_n + Pb$).

3. They cause losses of values, especially of silver, by volatilization.

4. They deteriorate the quality of associated metals (copper).

5. They, particularly arsenic, are detrimental to the health of the workmen.

Some of these objections are well taken, but some have lost much of their force consequent upon late discoveries and improvements in methods of ore treatment. Notice, for example, that the corrosive effect of molten arsenides upon iron and upon brick-work, being a result of chemical action which is now understood, shows the way to a ready means for decomposing such molten substances, as well as of resisting their prejudicial effects. Given molten metallic arsenides with access of air, and contact with siliceous material, and silicates of metals result. Pursuing the dependent train of reasoning toward its logical conclusion, and carrying out the processes indicated, we are led to an application of the pyritic smelting and bessemerizing principles, and experiments actually show that under the influence of the air-blast the arsenides are decomposed with ease, more readily in fact than the sulphides to which those principles have been heretofore adapted. Experiments made by the writer on mixtures of fused sulphides and arsenides show conclusively the greater facility with which the latter are decomposed, and how the elimination of arsenic takes place before that of the sulphur, and with what high heat it is accompanied.

We are here traversing new and unexplored ground, whereon one should tread circumspectly, and I have no desire to hazard predictions as to the outcome of the application of principles in fields to which they have not as yet been applied, nor to anticipate the results which others may be better entitled to announce; but it seems to me that we have in this habit of the arsenides, and probably of the antimonides, so amenable to the influences of the blast of air, a characteristic which will go far to offset their injurious behavior in other respects, and that will probably make their native minerals among the easiest of all substances to smelt.

9. The Genera of Matte Constituents.—We have, then, the following list of substances occurring in mattes:

1. Simple sulphides.
2. Sulpharsenides.
3. Sulphantimonides.
4. Arsenides.
5. Antimonides.
6. Magnetic oxide of iron.
7. Metallic iron, lead, copper, and gold.

More than fifty different compounds and simple substances have been reported at one time and another as existing in matte; but not one of them was essential to its existence as such; any one could have been removed and still leave matte. Matte, it would seem, must be made up of a mixture of definite chemical compounds, each of which has or may have its analogue in the mineralogical world. Double or multiple salts may exist among them, as is further suggested by the common phenomenon of the crystallization of even very complex mattes.

10. Wide Dissemination of Favorable Ores.—It is self-evident, and I need not argue the point, that we can always secure a mixture of ores which will on melting give rise to an effective carrier; for it is probably quite impossible to name, and almost as difficult to conceive of, a mining region where there are neither sulphur, arsenic, nor antimony; neither iron, lead, nor copper. From this consideration I think it may fairly be claimed that in this respect matting has an unimpeachable advantage over a rival

process which requires a large proportion of lead in order to become effective. This truth, important and far-reaching as it is, is of so obvious a nature that I am scarcely warranted in enlarging upon it. I prefer rather to leave it to the decision of those familiar with the conditions prevailing in our mining regions as to which process is likely to prove most generally practicable.

11. EXCESS OF MATTE FORMERS.—It is true that in matting we can usually obtain the necessary matte constituents with great facility; but we may and generally do have too much of them. That is to say: we often have ores which contain not only the heavy metals named, but also so much sulphur, arsenic, or antimony in combination, that it becomes necessary to get rid of the excess above the proportion needed for forming the carrier. It is then that the resources and skill of the metallurgist are most heavily taxed. We may get rid of the excess in either of two ways: the one, outside the furnace, by roasting as if preparatory to lead smelting; the other inside the blast furnace, when the operation is called pyritic smelting, being the latest developed and most interesting branch of matting.

It is the useful peculiarity of matte smelting, that we need be at no pains to cause the sulphur, the arsenic and the antimony to combine with the various valuable metals which may be present; for those obliging elements always unite of their own volition with the most intrinsically valuable metal, not directly, perhaps, in every case, but if not, the absorbing it by means of directly formed mattes or speisses. Provided that a sufficient number of matte-forming substances are present, all the values are certain to be saved, excepting of course, the unavoidable and usually small losses of the operation.

12. CONDITIONS GOVERNING THE ABSORPTION OF METALS. —The useful result of the matting fusion in the presence of sulphur and arsenic is the saving of the valuable metals about in this order, beginning with that one which is found to be extracted most completely: Gold, copper, nickel, cobalt, silver, lead. These, with iron, which is always present, constitute the metallic portion of the matte. The iron appears to be present only to take up whatever excess of metalloid there may be, its percentage diminishing with the increase of the other metals brought down. Excepting lead, neither of the metals named are crowded out of the matte by iron, but on the contrary that metal itself is prevented by them from combining with sulphur and arsenic under

the conditions prevailing in the furnace. This is fortunate; for not only are copper, nickel, and cobalt valuable in themselves, but the mattes in which they are contained are probably more generally efficacious in extracting the precious metals, than the matte composed only of sulphide of iron.

13. SUMMARY OF THE RESULTS OF MATTING.—It follows from the foregoing considerations that matting, as the result of several more or less tangible reactions, saves each one of the several valuable metals at one operation, concentrating them in one or sometimes two substances, of varying composition, often almost as complex as the ores from which they were derived. Again, as a corollary, if we diminish the relative proportion of matte formed, as compared with the amount of ore smelted, the composition of the product will be changed. We shall find less and less iron in it as we concentrate more, and also less sulphur; but we shall discover an increasing percentage of gold, silver, copper, cobalt and nickel, and to a certain extent, lead. In the common case of treating sulphur-bearing gold, silver and copper ores, if we keep on decreasing the proportion of matte (which in practice we effect by diminishing the proportion of available sulphur in the charge) we arrive eventually at a copper-silver-gold alloy, having passed successively through the various stages of copper-matte known as coarse metal, blue metal, white metal, and perhaps pimple metal, the important members of the copper-matte series, and beyond which, as we proceed to higher degrees of concentration, we pass from the domain of matte-smelting into that of metallic copper smelting, of which we have examples in Arizona and elsewhere.

While noticing the divergence of the two processes at this point, let it be added in correction of a misapprehension that seems to have arisen in some minds, that the copper-silver-gold alloy which has been mentioned, if made in a blast furnace, would be called black copper; if produced in a reverberatory, blister copper; and that the two products differ in important particulars.

14. PHENOMENA RESPECTING GOLD AND SILVER.—The experiments of Mr. Pearce (*Trans. A. I. M. E.*, XVIII., p. 454) demonstrate that plain ferrous sulphide exercises no solvent action on pure gold, the latter when melted with such a matte becoming diffused irregularly through it in the form of globules; and this finding is well supported by the matting experiments at the Boston Institute of Technology, recounted by Mr. Spilsbury (*Trans. A. I.*

M. E., X V., p. 767), where it was shown that the gold of pyrites con-
centrations was not absorbed in the matte arising from their
fusion. This evidence is conclusive as to the lack of absorbing power
of ferrous matte toward gold alone; and the further experiments of
Mr. Pearce upon the effects of the same kind of matte upon that
metal in the presence of silver are hardly less so. In brief, he
finds that an alloy of gold and silver is formed, which is dissolved
in the matte instead of being physically disseminated in it. One
would think that the solution of the alloy in the matte would be
similar to the solution of similar alloys in lead bullion; and from
this point of view the problem of saving the precious metals would
resolve itself into the saving of the matte and separating it from
the other products of fusion. As corroborative evidence of the
efficiency of ferrous matte as an absorbent of gold in the presence
of silver I shall cite the work done at the Toston and Deadwood
matting plants, where many thousand tons of pyritic ores have
been successfully treated with the production of such a matte.
Unfortunately, the details of their work have not been published
as fully as could be wished ; but if we may accept the assurances
of the managers, which I for one do not doubt, the extraction in
each case has been exceptionally complete. However, as bearing
on the technical point at issue, the evidence supplied by their con-
tinued practical success is not as satisfactory as could be wished,
for analyses are lacking to show that the matte is really entirely
free from other metals that might exercise an independent absorbing effect that cannot be disregarded. I think that the tendency
of what evidence we have is to prove the important effect of
relatively small quantities of elements: of stray and unconsidered
substances, as it were. Chemical analysis shows the extreme
complexity of what have been thought pure ores. Iron pyrites
often contains half a dozen of impurities which might have an
important chemical effect upon gold, while in the products of the
smelting furnace we find concentrated a surprising number of well-
known and even some rare elements; and as to those mattes which
have been reported as pure iron sulphide, it is very much to be
doubted that they do not contain at least a small proportion of
hitherto unnoticed substances; for it would be remarkable that a
deposit of ore of such character as to give rise to an absolutely
pure matte were found to exist. The influences particularly of
bismuth, arsenic and tellurium on gold contained in mattes may

be anticipated in advance of experiment, and form a fertile subject of study and deduction.

15. THE INFLUENCE OF COPPER.—The influence of copper, which plays no part in the experiments cited, is of great importance in practical work throughout almost the entire domain of matte smelting. It has been deemed essential to the perfect extraction of the precious metals that a proportion of copper should be present in the charge, and it is difficult to make some individuals believe that practical work can be carried on without it. It may well be that under some circumstances that metal is imperatively necessary to a good extraction; but that under certain other circumstances good work can be done in its absence there is no doubt. I need cite no evidence save that afforded by the two plants mentioned to prove the point, nor will I venture upon a generalization of my own concerning the conditions under which it is safe to rely upon a non-cupriferous matte for the extraction of both the precious metals; but I feel under the necessity of quoting the language of an eminent metallurgist whose name I am unable to use, who sums up tersely the belief to which he has come through years of experience: "Copper is not necessary in mattes which accompany acid slags; with basic slags it may be necessary."

The first clause of this dictum is amply proved by the gentleman's own work in producing such matte; and the second I find to agree with such evidence as I have been able to examine. I conjecture that there may be substances possessing as great or greater an influence than copper; but in their absence I regard that metal as essential to a clean extraction in all the cases which have come under my notice, where basic or even slightly acid slags were being produced. The beneficial effect seems to increase up to an uncertain limit with the percentage of the copper, for I have repeatedly found that a very low-grade copper matte failed to extract as high a proportion of silver as a matte richer in copper. I presume the extraction of gold too would be influenced in the same manner, but of this I have no evidence to offer. It was noticed at Anaconda, however, that mattes of a certain comparatively high tenor in copper left more silver in the slags, the expression there being that "the copper crowded the silver out of the matte." This observation is not inconsistent with the preceding, inasmuch as the Anaconda mattes are of much higher tenor in copper than those used mainly for the extraction of the precious metals, and it

may be that the very rich mattes do not possess the extractive
powers of the poorer ones. Besides, it is generally considered that
silver exists in such mattes as an integral part thereof, combined
with the sulphur, and so far analogous to the native copper-silver
sulphides; and bearing in mind the greater affinity of sulphur for
copper than for silver, it could hardly be expected that the latter
would be taken up by the metalloid to the exclusion of the former.
We may consider the condition of things as they apparently exist
in a reverberatory matting furnace toward the end of the smelting
operation, when the boiling has ceased and the excess of sulphur
has passed away; or, more to the purpose, that prevailing during
the concentration of argentiferous matte, when sulphur is being
eliminated and iron and copper successively oxidized. Should not
the silver, which is held, as Fournet teaches, by a tenure weaker
than the two base metals, be removed from the matte before the
others, did it there exist only as a sulphide intermixed with other
sulphides? But on the contrary it outlasts the iron and the sulphur and eventually separates as a copper-silver alloy. Gold is
somewhat more prone to leave the matte, but it too invariably does
so in combination with some other metal, and we do not often
hear of either of the precious metals having been seen in their
native form in or about any smelting-furnace product. These considerations appear to indicate that aside from their existence as
definite chemical compounds in matte, which we cannot help ad-
mitting, the precious metals possess other modes of combination
and diffusion, as I have previously indicated.

16. THE CARRIER PRACTICALLY CONSIDERED.—There are some
practical considerations connected with the formation of matte
which are of more immediate importance than the theoretical
points heretofore advanced. These relate either to the behavior
of the product in the smelting furnace, or to the subsequent operations of refining and parting. Generally speaking, the ferrous
matte would prove most favorable to the proper running of the
furnace, owing in part to its beneficial effect upon the hearth.
If, however, such a matte were produced in very large quantities,
special arrangements would be necessary to obviate its corrosive
effects. Under some circumstances we might find the arsenide
more advantageous than the sulphide mattes, on account of their
greater specific gravity, which allows them to settle out of
slags with more readiness. They might, for example, be profitably used in connection with a slag that runs high in zinc, which

often interferes prejudicially with the separation of matte; or
when the slag is extremely heavy from a preponderance of iron or
baryta. This point I will discuss in another connection.

If we consider the subsequent treatment of the carrier for the
extraction of the precious metals, pure iron matte would generally
prove most convenient, its refining being quite within the range of
simple and inexpensive processes and plants. In fact I may say
that plain ferrous sulphide is as easily treated as any ordinary
sulphureted gold or silver ore, and for all practical purposes may
be regarded as a simple ore. For its beneficiation we have a choice
of several methods: 1. Re-smelting, raw or roasted, with lead
ores; 2. Lixiviation with chemical solutions; 3. Amalgamation
with quicksilver in pans or barrels. More complex products
require, as a rule, more complicated and difficult treatment.
Such extractive processes, whatever be their nature, are almost
invariably preceded by calcination of the mattes—an operation
which is made much more difficult by their complexity. I have
indicated in the accompanying table the refining methods which
are made use of in the several cases cited. Obviously the choice
depends upon the surrounding conditions. If, for example, we
have the use of an amalgamating mill close by our furnace, our
endeavor will be to adapt that process to our wants. The successive lixiviation of the several valuable metals by means of chemical
solutions is practiced to a great extent, and affords peculiar advantages toward the attainment of pure end products. Various
methods whereby the valuable metals are sought to be extracted
from the molten matte or speiss have been proposed, and some
have been adopted with great success. The most important of
these is the Manhès process of bessemerizing—a method which
is of vast and increasing importance in copper metallurgy, and is
being rapidly extended to other metals.

17. TREATMENT OF MOLTEN MATTES.—Davies successfully de-
silverizes melted arsenide mattes (*Engineering and Mining Journal*
(?) 1888) by means of melted lead added to a bath of the matte,
kept in ebullition by means of an air blast applied as in the Manhès
process. Probert proposes the use either of lead or litharge, the
latter of which is decomposed by the materials of the bath, while
the mixing, which appears to be essential to these processes,
is effected in Probert's by the ascending bubbles of carbon dioxide
from lime carbonate in the lining of the containing vessel. That
these methods are imperfect, extracting only a moderate percen-

tage of the gold and silver values, is not particularly prejudicial, inasmuch as provision for returning the material to the furnace always exists. Heretofore the matte has been allowed to cool and solidify before being returned to the smelting furnace; but the writer has proposed its return while still molten, whereby its sensible heat is utilized and other important advantages gained. This process, which is applicable more particularly to the pyritic furnace, will be adverted to again in the proper connection.

It would seem that the Davies and Probert processes, which have proved useful in the treatment of arsenide mattes containing gold and silver, have not been practically applied to the treatment of sulphide mattes, although there would not appear to be any particular reason why they should not prove equally as advantageous therein. It is the opinion of those who have studied them that they will be found applicable to such mattes, in which case their sphere of usefulness would be much increased. On the other hand, the Manhès process, which has effected such wonderful results in the domain of copper extraction, has not within the writer's knowledge been practically applied in the decomposition of the arsenide mattes, which alone are the objects of the two other processes. However, the arsenides are broken up by the bessemerizing blast with the greatest facility, and their oxidation and concentration is, as far as the writer's experimental trials show, one of the easiest operations that confront the metallurgist.

The most interesting of the arsenide mattes are those containing cobalt and nickel, metals which have a strong affinity for arsenic— an affinity which is taken advantage of sometimes in the beneficiation of their ores when these metals are sought in the presence of substances which exercise an opposing influence. It has been found advisable under some circumstances to make such an addition of arsenic-bearing materials to cobalt or nickel ore as serves to bring about the formation of nickel or cobalt arsenide, while other heavy metals in the mixture separate therefrom as sulphides. In this manner it is possible to effect a useful separation of the two, even from very complex and difficult combinations. The nickel and cobalt arsenides are still very impure, and comparatively troublesome to deal with, requiring a prolonged succession of operations to obtain the metals in reguline form. The metallurgy of nickel has undergone great improvements within late years, consequent upon the increased demand for that metal. Lately the German system of matting has been applied to its extraction from

METHODS OF MATTE REFINING.

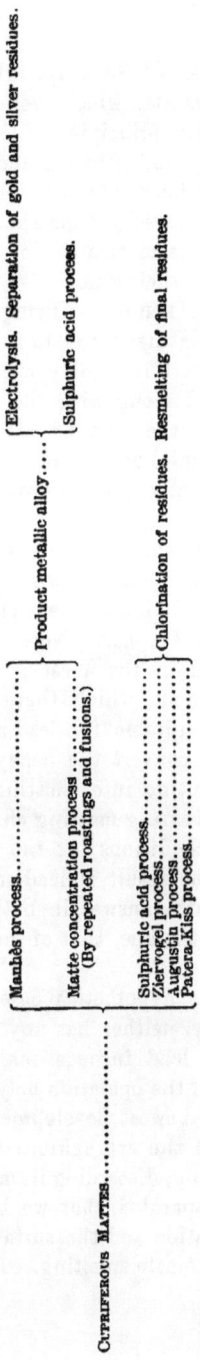

CUPRIFEROUS MATTE......

- Manhès process........ Product metallic alloy..... { Electrolysis. Separation of gold and silver residues. / Sulphuric acid process.
- Matte concentration process .... (By repeated roastings and fusions.)
- Sulphuric acid process......
- Ziervogel process......
- Augustin process......
- Patera-Kiss process.   } Chlorination of residues. Resmelting of final residues.

Process for production of nickel oxide. (Orford process, Bessemer process, Reverberatory process, etc.)

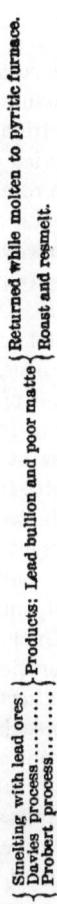

NON-CUPRIFEROUS MATTE........

- Roasted........
  - Amalgamation in pans or barrels.
  - Lixiviation ...... { Patera-Kiss process...... / Augustin process...... / Russell process........ } Chlorination of gold-bearing residues. Resmelting of final residues.
  - Chlorination.
  - Smelting with arsenic-bearing ores for nickel or cobalt arsenide.
  - Smelting with lead ores.
- Unroasted........
  - Smelting with lead ores. / Davies process........ / Probert process........ } Products: Lead bullion and poor matte { Returned while molten to pyritic furnace. / Roast and resmelt.

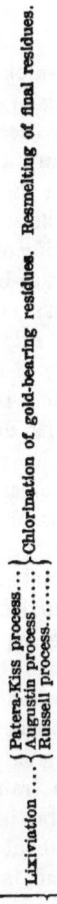

27

the roasted copper-nickel-iron sulphides of Sudbury, where the
product of smelting is copper-nickel matte, whose average composition is given in the table. As for the principles of the art as
carried on at Sudbury there is nothing novel; but grand results
are reached through the skillful application of old and well-known
means. Much of the success arises undoubtedly from the extraor-
dinarily rapid driving of the blast furnaces. which smelt twice
the average amount to be expected from their size.

18. COINCIDENCES OF SMELTING METHODS.—Owing to the
usual presence of sulphur in the mixture fused in lead blast
furnaces, a quantity of matte is the almost invariable accompaniment of the operation, it being produced along with the metallic
lead, in proportions depending upon the amount of sulphur
present. Should the percentage of sulphur be increased, the
matte-fall is also increased, until eventually the lead quite fails of
reduction and enters the matte, as do the gold and silver, when
present in the charge. Instead of lead smelting, the operation
has now become matting of the German system, which can be
and sometimes is carried on in furnaces similar to or even identical
with the prevailing type of lead blast furnace. Note that the
operation of lead smelting as carried on in the West is really a
mixture or combination of the two methods, which therein over-
lap each other. In smelting a charge suitable for lead smelting
except as containing sulphur enough to convert the heavy metals
into sulphides, we transform lead smelting into matting. And
conversely, by abstracting from a lead-bearing matting charge the
sulphur, we bring about the necessary conditions for the production of metallic lead, and in so far the result is lead smelting.
The same furnace and accessory apparatus answer in both cases.
The production of metallic lead marks the one, that of matte the
other process.

19. CONVENTIONAL IDEAS.—In these hypothetical cases, where
the two methods approach so closely, neither has any decided
advantages over the other. Matting a lead furnace charge plus
sulphur is easily done, but the utility of the operation may be but
slight. It is not matte smelting in its newest developments, nor
are any of the more valuable features of the art exhibited. It is
only when we get away from lead smelting, discarding its mixtures,
its slags, and to a large extent its apparatus, that we begin to
realize the extent and variety of application and the suitability in
diverse circumstances that characterize matte smelting. It is for

this reason if for no other, that the application of the methods of
lead smelting to the art have not been found to succeed, and why
metallurgists of that school have not been and are not likely to
be successful in their experimenting.

Speaking from the experience gathered in several years of experiment and research, I must say that it appears to me there is
no possibility of successful matte smelting when following the
lines laid down by the lead smelters. In fact it is only when we
leave the beaten track and strike out upon a road of our own that
we put ourselves in the way of successful achievement; and
furthermore it appears at present that the most noteworthy
successes have been and are destined to be reached by going in
many respects contrary to tradition and to ordinary practice.
There are very few occasions, and I think none, where a charge
suited for the lead furnace can be more profitably matted. The
great and manifold advantages of matting do not appear at such
times, but principally when there is stress in procuring lead or
fluxes or particular descriptions of fuel. Several instances are on
record where results deemed satisfactory have been achieved by
smelters who have run down charges in the lead furnace, inten-
tionally producing matte instead of lead bullion: but it is difficult
to see in any such accounts, however complacently told, the in-
dication of a proof that any advantages were gained, and there
must have been some lost. Clearly there is no object in making
matte when we can with more profit make bullion.

Certain smelting men, identified perhaps with lead extraction,
and very likely impressed with the perfection to which that art
has attained of late, have endeavored to extend the principles and
practice thus acquired to matting gold and silver ores. Their
success and the results of their work can be summed up by saying
that they appear to have achieved a gratifying measure of success
on very easy lines of effort, smelting, for example, a great deal of
old lead slags along with a little ore, and getting the gold and
silver in the form of matte. Such results, satisfactory as they may
be from a business point of view, throw little light upon the important questions of ore treatment, and only seem to confirm the
position of those who maintain the erroneous yet not uncommon
view that matting is inferior and subsidiary to lead smelting.
There are those whose position in the world of metallurgy should
forbid it, still perpetuating the error of thus subordinating matting, for which they patronizingly predict a career of usefulness,

limited probably, in connection with the other branch of smelting.
But matte smelting stands alone. It is not connected necessarily
either in principles or in uses with copper or lead smelting or any
other branch of metal production. The field of its application is
vastly wider than theirs, and in its adaptability to diverse conditions no other branch of metallurgy can ever be expected to rival it.

20. RELATIONS WITH LEAD SMELTING.—Returning from this
digression to the subject in hand, it occurs to me to say, that if
matting a lead-smelting charge, plus sulphur be an easy thing, it
does not follow that lead-smelting a lead-bearing matting charge
minus sulphur is always an easy, an economical, or even a possible
thing. For not only may we have present such metals as copper,
nickel, and cobalt, which it is desirable to recover, but our ores may
be of so acid a composition that the loss of lead by scorification as
silicate, and by volatilization on account of the high temperatures
necessary to form acid slags, would make lead smelting ineffective
at least. We are consequently debarred from the use of lead
smelting, first, in the frequent cases where valuable metals other
than lead, silver, and gold are to be won; second, where sulphur
is a constituent; and, third, where the resulting slag would contain
more than 38 per cent., or thereabouts, of silica. Casually read,
my " second " might be excepted to by those familiar with the
character of the ore treated by lead smelters; but it should be
borne in mind that lead blast furnaces are engaged, as a rule, in
making matte as well as lead bullion, and are equally as well calculated for the one as for the other. The extraction of the lead
by that method entails the almost equally perfect saving of the
sulphur, with the production of matte, which requires additional
processes for its reduction. These processes have become so
recognizedly a part of the duty of lead plants that the term lead
smelting now signifies the production and reduction of mattes as
fully as the production of lead bullion. Blast-furnace mattes,
whether produced alone (German system) or in conjunction with
lead bullion (lead smelting), follow frequently enough the same
course of reduction, namely: oxidizing roastings alternated with
re-smelting with lead-bearing substances, producing an alloy of
lead with gold and silver (bullion), and a concentrated matte containing copper, an perhaps cobalt and nickel, with small quantities of the precious metals. The presence of sulphur, therefore,
which makes the theoretical difference of the processes, entails in

each case the requirement of the decomposition of the resulting matte.

Since the inception of the Austin process, and incited by its success, many ideas of novelty and importance have been evolved by those working on parallel lines of inquiry, inventions are multiplying, and research, experiment, and practice go hand in hand. The subject presents itself in a much broadened aspect. No longer from the narrow standpoint of the "process man" can we embrace the unbounded prospect that lies before us. No longer is it tolerable to regard pyritic smelting as the exclusive domain of a single great virtuoso, but rather as the property of those who can master and add to it.

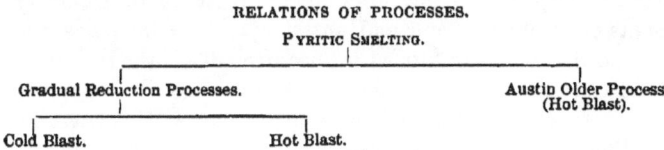

RELATIONS OF PROCESSES.
PYRITIC SMELTING.

Gradual Reduction Processes.	Austin Older Process (Hot Blast).

Cold Blast.	Hot Blast.

DEFINITIONS.

Gradual Reduction System.—The application of oxidizing currents of hot or cold air to ore mixtures fed in the ordinary way (layer feeding). Examples: Bartlett Works,* Canyon City, Colorado; Porphyrite Works, Mineral, Idaho; Bi-Metallic Works, Leadville, Colorado.

Austin Older Process.—The application of oxidizing currents of hot air to unmixed ores. Example : Toston, Montana.

21. RELATIVE PROPORTIONS OF CHARGE AND PRODUCT.—Before considering the relative weights of material charged and matte or bullion produced in furnace practice, it will be useful to contemplate the proportion of valuable metals in the material which it is proposed to treat. In the accompanying table will be found, under the appropriate heads, some information on this point. We notice that the proportion of the metals to be saved varies widely in different cases. For example, the copper in the concentrates at Butte forms about one-fifth of the charge of the reverberatories. The lead and silver in the customary charge at Tacoma aggregate

* Mr. Bartlett's efforts have been directed mainly to the treatment of mixed ores of zinc, lead, and copper, with silver and gold, recovering zinc-lead pigment and copper-gold-silver matte. For description see the *Engineering and Mining Journal,* Vol. LVI., pp. 3, 366, 594.

one-eighth; the copper and silver at Mineral from one-fortieth to one-sixty-fifth; the silver and gold, which made up the sole values at Toston, together constituted but one twenty-two-hundredth part of the mass of the charge. Further inspection of the data will reveal the like interesting facts concerning the work at other establishments. It is evident that the necessities of the case when there are certain proportions of copper, nickel, lead or other weighty substances to extract will govern the percentage of product, which is therefore uncontrollable in so far. Again, having supposedly to drench the slags with a vast proportion of matte when we work only for gold and silver, furnishes a second instance of the practical requirements of matting. And finally come those necessities imposed on our work by the presence of significant amounts of sulphur, arsenic and antimony, whose influence upon the proportion of matte formed is uncontrollable in the lead-smelting and German systems of reduction, but are partially controllable in the pyritic system, whose influence thereupon I will later describe.

22. PROPORTION OF VALUES.—From other columns of the table may be derived the proportions of the valuable metals to the whole weight of product. At Tacoma and all other lead smelters the bullion product contains practically 100 per cent. of valuable metals, or its whole weight. At Mineral it is from 15 to 41 per cent.; at Butte and Anaconda about 50 per cent.; at Sudbury, 42 per cent.; at Toston, slightly over four-tenths per cent. The purpose of this is to show indirectly what an overpowering proportion of practically worthless iron, sulphur, antimony, etc., it has been found necessary in many cases of matte-smelting to extract and subsequently to still further reduce preparatory to the removal of the valuable metals which we design to win. In these respects we may look for improvements in matting practice in two directions. First, in the diminution of the proportion of matte produced to gold and silver or other valuable substances saved in the treatment of appropriate ores; and second, in the possible utilization under favorable circumstances of some of the accompanying elements which are now deemed worthless or even prejudicial.

That important improvements will take place in the latter regard there seems no reason to doubt, for, as indicated in the introductory remarks, it seems that no insuperable obstacles exist to prevent the direct manufacture of sulphuric acid from the gases evolved in matting, especially, as will be pointed out, in that form known as pyritic smelting. A great deal might be said at this time and in this connection concerning the utilization of the several now valueless elements which enter the matte in precious metal smelting, but as such remarks would be in large measure anticipatory of practical results, I will at present refrain.

23. POSSIBILITIES OF CONCENTRATION.—Regarding the proportion of matte necessary to bring down the gold and silver of such charges as I have worked, I may say I have not obtained results of a satisfactory definiteness, nor am I aware that the minimum proportion is known to have been attained in any case. I have used experimentally as small a proportion as two and a half per cent. of matte, as reckoned on the charge, while the average in my regular work has been five to seven per cent. I have also produced at times 20 per cent., or one-fifth of the weight of the materials charged. It might reasonably be supposed that conditions so diverse would give rise to somewhat varying degrees of extraction of the values; but the experience gathered may be summed up by saying that the savings were as high in one case as in another, the like conditions prevailing as to temperature, and especially as to the composition of the products. It seemed to me when producing clean slags by the aid of a given proportion of matte, that I could still get the same clean slags with a less quantity of matte. To what extent it would be possible to reduce the matte production and continue to do good work it is difficult to say, although it would evidently depend on the chemical character of the slag formed, and doubtless, within certain limits, of that of the matte. A friend, in whom I have great confidence, tells me that his rate of matte production is 5 per cent., and that his slags are as well cleaned as with a higher ratio. The evidence, therefore, is to the effect that a very great degree of concentration is practicable; greater in fact than has been generally known, but additional practice is necessary to set bounds thereto.

24. PRACTICAL ADVANTAGES.—This very important feature, which is not enough known or studied, constitutes one of the greatest claims to usefulness of this branch of metallurgy, especially in its application to gold and silver extraction. It is obvious that the remark is not applicable to the matter of the saving of copper, lead or any other substance which forms a considerable part of the weight of the ore. Scientific students of the art will on reflection comprehend the full bearing and importance of the principle, as a property of matting which makes it of the greatest

utility in the treatment of gold and silver ores in regions particularly remote, where the costs of transportation tell heavily as against
the production and shipment of weighty products, such, for
example, as lead bullion.

25. CONDITIONS GOVERNING CONCENTRATION.—When put-
ting forty tons of charge into one ton of matte, in one operation,
I had not necessarily reached the utmost limit of concentration;
on the contrary I think that even that high rate might under
appropriate circumstances be profitably exceeded.   But practical
difficulties intervene; the minute amount of matte produced may,
by faulty manipulation of the furnace, be stopped altogether; or,
on the other hand it may by other agencies be unduly increased.
The most delicate and skillful handling, far beyond anything
required or practiced in ordinary smelting, is essential to maintain
the regularity of the matte-fall at such minute proportions.   Thus
the mechanical difficulties of smelting interfere to prevent the
full attainment of those great advantages which are found to flow
from the very highest concentration.   On the whole, under ordinary
conditions, I do not regard it as advisable to seek a higher rate
than twenty-five into one, and this of course only when dealing
with gold and silver ores.   I need hardly remark that copper
matting allows no such rate of concentration, the whole aspect of
the problem being different, nor need I enlarge upon the conditions
under which our lead smelting is carried on, as to the employment
of a percentage of lead which must not fall below eight or thereabouts, but generally reaches twelve.   I do not doubt that the fair-
minded metallurgist has already conceded those advantages which
advocates of the matting processes claim in this direction.*

There are no matting plants running in America, and probably
none in the world, under conditions which make such an extreme
degree of concentration advisable or necessary, and therefore we
cannot point to examples which would illustrate to the full the
advantages which flow from the very highest rates of concentration.
We can imagine, however, the not uncommon conditions under
which such high rates would possess the maximum of advantages,
and they are in many respects similar to those prevailing in the example cited.   The most stringent conditions, such, for example, as
those which prevail in many mining regions in Mexico, while in

* Compare Kerl, "*Metallurgy of Silver*," who asserts that the raw smelting
of silver ores should be accompanied by the production of from 30 to 50 per
cent. of matte, thus concentrating two to three into one.

many respects unpropitious to smelting of any sort, may not be fatal to matting when it is carried on with a view to this highest concentration of product.

26. SPECIFIC GRAVITY OF MATTES. — Upon this important subject I will venture but few remarks, first referring the reader to the appropriate text-books for information as to the specific gravity of the simple substances which compose mattes. It would seem that although the lightest sulphides which enter into a mixture may not, as in the cases of calcium, zinc, and manganese sulphides, exceed a specific gravity of 4 or thereabouts, the heaviest of the arsenides reach a greater weight than cast iron, and even approach nearly to the gravity of copper, which is above 8.5. Their specific gravities are such that we can arrange the various compounds in three groups having well-marked differences in several respects, but differing mainly in gravity. The arrangement is as follows :

Group 1. (Substances having a specific gravity not greater than 4.7.) The sulphides of zinc, molybdenum, calcium, and manganese.

Group 2. (Specific gravity between 4.7 and 5.5.) The sulphides of barium, iron, cadmium, nickel, cobalt, and copper, and the magnetic oxide of iron.

Group 3. (Specific gravities ranging from 6 to 9.) The sulphides of silver, lead and bismuth; the arsenides and antimonides, and the sulpharsenides and sulphantimonides of silver, copper, bismuth, lead, iron, cobalt, and nickel, and metallic lead, iron, and copper.

The intermixture of these compounds necessarily produces a matte of intermediate specific gravity, or it may produce two or even three substances of varying gravity, which separate in the hearth of the furnace. I have prepared the annexed table (see p. 36) for the purpose of showing graphically the influence of various elements upon the specific gravity of mattes, and enabling the student to deduce the relative density of any proposed product. I need hardly remark that the matter is of great practical importance in its bearing upon the separation of the matte from the refuse materials in smelting, and will repay close study and attention, but the results obtainable by use of the chart are only to be considered approximative.

27. EXAMPLES.—The commonest description of matte which is produced by the lead smelters would probably be of about the fol-

lowing composition: Lead, 15 per cent.; copper, 6 per cent.; sulphur, 23 per cent., the remainder mostly iron, with small quantities of zinc and arsenic, and minute amounts of a dozen or more of subordinate elements. Experiment shows that the specific gravity of such a matte is about 5.3. An increased proportion of zinc lightens it very much, and increased iron lightens it some-

SPECIFIC GRAVITIES OF THE MATTE FORMERS.

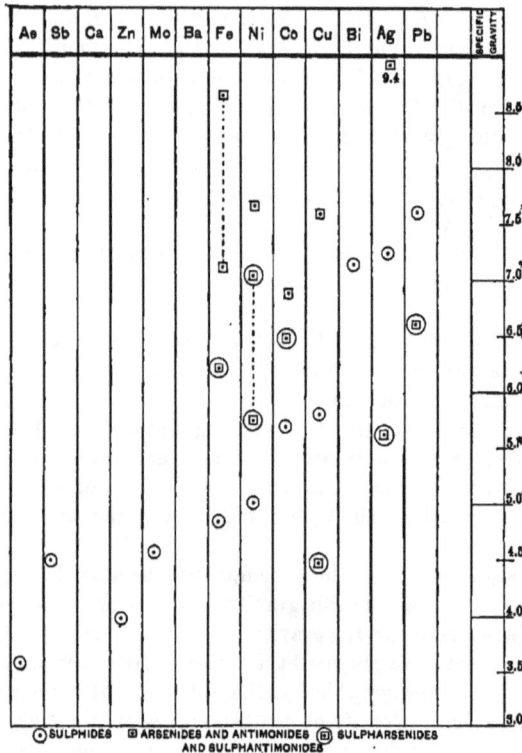

As	Sb	Ca	Zn	Mo	Ba	Fe	Ni	Co	Cu	Bi	Ag	Pb	SPECIFIC GRAVITY

⊙ SULPHIDES   ▣ ARSENIDES AND ANTIMONIDES   ⊚ SULPHARSENIDES AND SULPHANTIMONIDES

what, but in a less degree. Increased copper, and especially lead, add to the gravity; and obviously the substitution of arsenic or antimony for the sulphur also renders it more dense. The heaviest matte with which I have had practical experience is an antimonial substance containing 25 per cent. of lead and 20 per cent. of copper, and whose specific gravity was 8.07. Accompanying it were small quantities of a still weightier matte having a gravity of 8.3, but the composition of which I did not ascertain.

SECTION TWO.

THE SLAGS.

28. PHYSICAL CHARACTERISTICS OF SLAGS.—Any sort of slag
that will melt at all may be used in one or the other department
of matte smelting, notwithstanding any peculiarities of its chemical constitution; but we are compelled to attend none the less
closely to these peculiarities when we seek to do thorough or economical work, and above all when we would rival the best performances of the scientific lead smelter. Slags have been and are
being made in practice which contain as much as 65 per cent. of
silica; or 40 of lime; or 23 of alumina; and manifesting what
appear to be the most abnormal qualities. Some slags have the
color and texture of stoneware; others are black, and very heavy
and brittle; some slack and fall to pieces like lime; others with-
stand the heaviest blows of the sledge-hammer. Some are thick
with unmelted fragments of quartz, like raisins in a pudding—
such cannot be handled in the blast furnace at all; while others of
more favorable composition fuse easily and drive fast. Some run
beautifully, keeping the furnace in good condition, while others
necessitate blowing out almost daily. Some contain nearly a fourth
part of alumina, a difficult constituent; while others are made up
largely of baryta and magnesia, which lead smelters abhor. Yet
all these slags, notwithstanding their peculiarities, serve their
especial purpose capitally, being made under conditions which
render them both economical and profitable. In discussing the
peculiarities of slags, it is common to characterize the use of some
kinds as "good practice," or "not good practice." We hear a
a great many such opinions advanced with no foundation other
than some preconceived idea of what is fitting in metallurgy.
Apparently, however, there is no real test of good or bad practice
in metallurgy except the financial test. It is good practice when
we make the most money; it is bad when we lose. Any particular
operation is good practice so far as it pays only. In this view the
examples of slags presented herewith are representative of the

"best practice," being made under conditions which rendered them practically the best that could be made.

29. RANGE OF PRACTICAL SLAGS. — Compare the diverse examples given in the Table of Work Done with those slags made in our best lead-smelting works, to gain a commensurate idea of the vastly wider scope and applicability of the matting processes. From these examples, the best attested which it is possible to procure, it will be seen what striking diversity is permissible in the composition of slags in this type of smelting. Thus we may have basic or sub-silicates; singulo-silicates, sesqui-silicates, bi-silicates and tri- or even quadri-silicates, when the atomicity of the bases will allow. And none of these, I may add, are incompatible with economical work. Experience shows that we may have in different cases:

Silica	from 25 to 70 per cent.	
Alumina	" 0 to 23	"
Ferrous oxide	" 0 to 70	"
Manganous oxide	" 0 to (?)	"
Lime	" 0 to 40	"
Magnesia	" 0 to 12	"
Zinc	" 0 to 22	"
Baryta	" 0 to 52	"

30. COMPARISONS WITH LEAD SLAGS.—We have therefore the largest liberty in choosing the slag for our projected operations, compared with which the lead smelter finds himself confined within very narrow limits. It has been laid down by one of the foremost lead smelters of the day that in order to do the best work the lead slags should not contain less of silica than 28 per cent., nor more than 37 per cent. Lime, he insists, should not go below 10 per cent., while the highest percentage of this base in any recognized slag-type is 28 per cent. They eschew magnesia entirely whenever possible, although one of our foremost metallurgists uses it extensively in matting and says he likes it as a flux.* Heavy spar, which has been hitherto a great bugbear to all blast furnace managers, is found now to be a valuable flux, almost as desirable as lime, excepting for its lower saturating power and greater

---

* Dr. Carpenter at his Deadwood works has been signally successful in adapting the German system to the treatment of highly siliceous gold ores, which he fluxes with magnesian limestone, producing very acid slags, whose bases are alkaline earths—a course of procedure to which this country affords no parallel.

specific gravity. It is very easily eliminated in the pyritic furnace, where with proper management it can, as I believe, be largely driven unchanged into the slag, mixing therewith, but without decomposing. Under other conditions of temperature, basicity of slag, etc., sulphuric tri-oxide is volatilized and the baryta unites with silica as in the reverberatory process. Still another result, but as a rule a less useful one, is brought about in the German process and in lead smelting, where heavy spar being reduced to sulphide of barium, by double decomposition with metallic bases, adds largely to the matte fall.*

31. THE MORE ACID SLAGS PREFERABLE.—As a consequence of our ability to handle the greater variety of slags, it follows that on the whole less flux is required than is found necessary in lead smelting, and, speaking in general, less slag is produced. We can matte a given quantity of average ores by the aid of less fluxes than we can lead-smelt them. The production of a less quantity of slag entails less expense for handling, as well as less loss from entrained† matte particles in the slag, and less heat carried away by it. That it is possible by skillful composition of charge, and watchful care in running, to make slags, especially acid ones, freer from the precious metals than are known in lead smelting, we have the word of more than one able metallurgist familiar with the practice of both methods. I believe that certain probably exceptional slags have been made in the matting furnace which are cleaner than have been reported in lead smelting, running as low as 30 cents per ton, or even less. But in general it is only safe to claim that under similar conditions the lead and matte slags will pratically assay the same in gold and silver. In other words, the conclusion is that there is no practical difference in the extractive effects of lead bullion and matte. The recognized types of lead slags may be quite as easily formed in the matting furnace, and where the nature of the ore mixture seemed to demand they might be profitably made. To make special efforts to achieve particular types of lead slag, however, would be impolitic under most cir-

---

* Kerl (*Metallhüttenkunde*) describes the intentional formation of matte by the reduction of heavy spar in connection with iron ores. See also *Mineral Resources*, 1874, p. 417.

† I take the liberty of using this word, taken in the sense in which it is used by mechanical engineers, because there seems to be a certain analogy between water carried by and included in steam and matte carried by and included in the slag.

cumstances, and a matter of doubtful utility in all. It should be
remembered that the recovery of metallic lead, which is the most
exacting requirement in lead smelting, plays no part in matting,
so that because of this dissimilarity of conditions the writer has
discarded lead slags in the matting furnace, finding in the more
acid ones, unknown to the lead smelter, a more promising subject
of experiment and research.

32. SPECIFIC GRAVITY OF SLAGS.—Those questions of specific
gravity which so profoundly affect the subject of matte-production
enter not less prominently into that of slags. In the effort to
maintain such differences in the densities of the concurrent products as will entail an adequate separation, we may take measures
to increase the gravity of the matte or diminish that of the slag.
The substances which enter into the composition of slags are principally the following, having a specific gravity (fused specimens.
mainly artificial), approximately of:

The singulo-silicates of iron, manganese, and zinc, about 4.

The bi-silicates of iron, manganese, and zinc, about 3.5.

The basic silicates of alumina, from 3.2 to 3.4.

The acid silicates of alumina, from 3 to 3.2.

The silicates of magnesia, from 3 to 3.3.

The silicates of lime, from 2.6 to 3.

The alkaline silicates, about 2.5.

Uncombined silica, 2.6.

The bi-silicate of baryta, 4.4.

The silicate of lead, 7.

Ferrous sulphide, 4.8.

Calcium sulphide, 4.

Magnetic oxide of iron, 5.

Sulphate of baryta, 4.5.

As regards these silicates the rule which governs specific gravity
is that it suffers a decrease with the increase of silica. Or in other
words the more acid the slag the lower its density, the same bases
remaining. According to this the tri-silicates should be still
lighter than those mentioned—a supposition which is borne out by
facts. Slags can be and very likely have been formed having a
gravity of less than 3, and possibly as low as 2.6, as mentioned by
Balling; but materials for such a formation are, and must continue, very scarce and impracticable. The highly siliceous slags
of Swansea (see table) have a density of 3.21, and rank among the
lightest which are made in the regular way of smelting.

SECTION THREE.

PYRITIC SMELTING.

33. THE CHOICE OF FUEL.—What is mainly sought in smelting is cheap heat. Those who contemplate matting operations, looking about for a source of heat, are quick to recognize the advantage which they have over the lead smelters in virtue of the very wide liberty of choice of fuel ; for while the latter are usually restricted to the use of the blast furnace, with coke or charcoal (and coal experimentally) as fuel, both blast and reverberatory furnaces are employed in matting, fired with wood, coal, charcoal, coke, or gas, according to circumstances. We have furthermore an important and very interesting source of heat from which the lead smelter is necessarily debarred, in that in pyritic smelting we are able to burn certain ores themselves as fuel. It is in its application to the treatment of highly sulphureted compounds that matte smelting has received its latest and widest development. Such substances as pyrite, chalcopyrite, pyrrhotite, arsenopyrite, etc., which in lead smelting are invariably roasted before fusion and in the ordinary form of matting frequently are, have been discovered to possess the most valuable properties as fuels, and through the efforts of American metallurgists have been brought into practical use as such in smelting. This process is a new one, having been in practical existence for but a very few years, although it is probable enough that its germs may have existed for a much longer period. I have no wish to forestall whatever may be said as to the history of the idea of pyritic smelting, but I am glad to be first to assert its standing as a distinct art, to define it, and also to place the credit of its inauguration as a distinctive process where it justly belongs. There can be no doubt that the honor of first putting pyritic smelting on a practically useful basis belongs to Mr. W. L. Austin, whose experiments at Toston, Montana, first demonstrated the utility of the hot blast in this line of ore reduction, and eventually paved the way to the successful introduction of the pyritic principle as an important feature of metallurgy. This language would appear to be entirely justifiable in view of the fact that plants on the pyritic system are now (June, 1894) in successful operation in direct competition with older methods of beneficiation, and that the process is being rapidly introduced in other localities. The pyritic system has thus attained along with an excellent measure of success a deserved and considerable celebrity which, naturally enough, lacks a good deal in discrimination, because attaching to methods by no means understood even by metallurgists, among whom not half a dozen in the United States have ever had an opportunity of becoming practically familiar with any one of the various pyritic methods. It is even imagined by many that only one pyritic method exists, and that it covers the whole field of pyritic smelting; but I hope to correct this misapprehension, the tendency of which is injurious to the progress of metallurgy, and show that abundance of unoccupied space is left for the efforts of other inventive spirits.

34. DEFINITIONS AND DISTINCTIONS.—In order to remove the uncertainties and misapprehensions connected with the term pyritic smelting, it should be restricted to that department of blast-furnace matting wherein a portion of the heat required for reduction and fusion comes from the oxidation of a part of the ore. Or, more briefly, wherein ore is burned for fuel. The term pyritic is not specially descriptive of any matting process; but since it has become familiar in metallurgy it may well be retained if properly restricted in its application. That it has not previously been so restricted every one who has been familiar with recent technical literature will admit.

35. PRIME DISTINCTION OF PYRITIC SMELTING. — It is, then, the prime distinction of pyritic smelting that ore itself is burned therein in the furnace. It has long been known that certain of the metallic sulphides which accompany or contain so large a proportion of our mineral wealth will, under favorable conditions of exposure to the atmosphere, absorb oxygen spontaneously, producing an elevation of temperature, and even incandescence, and become imperfectly oxidized. This tendency is taken advantage of in the operation of roasting these ores in furnaces, wherein the end sought is the same, but the process is hastened by applying a rapid air current, supplemented by heat, which is the familiar operation of roasting or calcining. The same tendency to absorb oxygen is the foundation of pyritic smelting, wherein the operation

is brought about in a more rapid way, with intenser heats and more powerful air currents, the net results aimed at being the roasting of a part of the sulphides, plus the smelting of the roasted material along with the unroasted part and the remaining gangue constituents of the charge—all in one operation. This very summary and effective treatment has two results bearing directly upon its metallurgical economy, namely: it obviates in so far the necessity for a preliminary roasting, and it reduces to an important extent the proportion of costly fuel required in the smelting.*

36. BASAL PYRITIC REACTIONS.—It will be useful at this point to call attention to some of the reactions which are presumed to go on inside a blast furnace. We are accustomed to the frequent use of the terms "oxidizing atmosphere" and "reducing atmosphere," both of which have important places in technical and scientific literature. Both these, in spite of their wide use, are indefinite, if not misleading, when applied to the conditions prevailing inside smelting furnaces. No furnace atmosphere can be unqualifiedly reducing or oxidizing in its effects. For example: in an iron blast furnace the atmosphere is oxidizing toward the fuel (carbon), and reducing toward the iron ores present. In the like manner, in the copper blast furnace, it is oxidizing toward the fuel and reducing toward the copper and iron oxides. In the lead furnace, while oxidizing the fuel, it tedns to reduce (using the word in its chemical sense) all the heavy metallic oxides present, and also the other oxidized compounds, including those of sulphur. In the German system of matting we find similar conditions prevailing, the oxidizing effects extending only to the fuel, while all the higher oxides are reduced, some suffering conversion into sulphides, the remainder into silicates. But in the pyritic system the oxidizing effects predominate. Not only is the carbonaceous fuel burned, but the excess of oxygen beyond what is required for that purpose enters into combination with the various metals and metalloids of the charge, according to the play of chemical affinity, producing oxides, some of which enter the slag, while others, more volatile, pass out with the smoke and spent gases. The blast, which in the iron and lead furnaces produced only a combustion incomplete and imperfect, of the coke and charcoal, in the pyritic

---

* The percentage of sulphur eliminated in various operations is approximately as follows : In roasting preparatory to lead and matte smelting, about 85 per cent.; in reverberatory smelting of ordinary charges, 13 per cent. (Vivian); in pyritic smelting, from 65 to 85 per cent.

furnace tends to burn more thoroughly the fuel, and more or less completely the oxidizable constituents of the ore also.   It may be said that the complement of the oxidation is the sulphidation; and the complement of the scorification of the heavy metals is the formation of matte.   Sulphur in the pyritic furnace performs functions entirely similar to those of carbon in the high iron furnace.   Carbon abstracts oxygen from oxides; by uniting with iron it forms cast-iron, a fusible substance; by uniting with oxygen it generates heat.   Sulphur abstracts oxygen from oxides; by uniting with the heavy metals it forms matte, a fusible compound; by uniting with oxygen it generates heat.   In these respects arsenic and antimony also act in a manner not dissimilar to sulphur, by their oxidation generating heat and carrying out the other reactions on which the process depends.

37. DEDUCTIONS FROM THE FOREGOING PRINCIPLES.—From the foregoing considerations it appears that the efficiency of pyritic over plain matting (German system) in any given case will be in proportion to the relative amount of oxidizable constituents in the ore charge; or, more strictly speaking, it will be proportionate to the absolute amount of heat units made available through the decomposition of the sulphides.   What the available heat from this source will be in any particular case depends on the extent to which oxidation is carried and upon the heat equivalents of the various sulphides, etc.   The latter may be computed from the heat equivalents of their elementary constituents, which are known. The problem in this aspect is identical with that referred to as "the heat balance of the blast furnace"—a subject which has been so profoundly treated by the iron smelters in its applicacation to their own pursuits.

38. MEANING OF THE TERMS OXIDATION, ROASTING, ETC.— We must not lose sight of the fact that it is to the oxidation of the heavy metals that may be in combination with the sulphur, the arsenic and the antimony that we are indebted for a large proportion of the heat evolved in pyritic smelting—a fact liable to be overlooked in the general view.   Pyritic smelting burns more than the sulphur; and roasting is more than desulphurizing.   Mr. Austin even attempts to show that the heat evolved by the burning of the iron is better for his purpose than that produced by the burning of the sulphur, the latter largely becoming latent in gaseous products, while that produced by the formation of ferrous

oxide remains therein during the critical period of the scorification of that substance.

As it is probable that no charge is ever smelted in a matting furnace without being indebted for a part of the heat to the oxidation of some sulphur or iron, however little this amount may be, it follows that the systems of blast-furnace matting are in this respect only different in the degree to which the oxidation is carried. In the German system this is so inconsiderable that it may be disregarded, considering the function of the sulphur, etc., to be only to form matte, while that of the blast is merely to oxidize carbon, whereby the necessary heat is evolved. It would appear to some that as a necessary corollary to this proposition the only difference between the two rests upon the amounts of air blown in; and that to mutually transform the German and the pyritic systems we need only vary the power of our blowing engines. And I may add that it is upon this assumption that a good many experimental trials of the latter system have been carried on. That the assumption is erroneous and that failure inevitably followed the experiments are equally certain.*

39. EXTENT TO WHICH OXIDATION MAY BE CARRIED.—As a necessary result of the fact that the atmosphere in the pyritic furnace is so far oxidizing as to burn sulphur and iron, it follows that it will also burn all the combustible compounds of carbon and hydrogen which may be present, including carbonic oxide, the hydro-carbons, and also, presumably, free hydrogen. These substances, let it be noticed, are among the usual products of the imperfect combustion which takes place in iron and other furnaces, where a reducing action has to be maintained. The ordinary course of blast-furnace smelting gives rise to gaseous products which are susceptible of being still further oxidized with the generation of much heat. The pyritic system, producing fully oxidized gases, lays claim to a more complete utilization of the

* The principal heat equivalents with which we have to deal are : Coke and charcoal, about 8,000 units ; olefiant gas, 12,000 ; marsh gas, 13,000 ; carbonic oxide, 2,400 ; sulphur, 2,300 ; iron, 1,576 ; zinc, 1,800. The assumption that the calorific efficiencies of different substances in the blast furnace are in direct proportion to their heat equivalents is not apparently borne out by observation. For example, the calorific efficiency of metallic iron should be about one-fifth that of carbon ; but experience shows that it is much higher. We have, however, to consider the latent and specific heats of the resulting compounds, and also the extent of the oxidation effected. On the smelting efficiency of metallic iron, see Raymond, *Mineral Resources*, 1870, p. 443.

fuel. There passes from an iron blast furnace a great deal of inflammable gas; the lead, copper, and German matting furnaces each give off gases in considerable quantities which contain unoxidized constituents possessing a high heating power; but the gaseous products of the pyritic furnace, like those of the reverberatory, should be so oxidized as to be incapable of further heat-producing reactions. In this connection I may mention that sulphur, which is volatilized so largely, appears to me to be converted partly into sulphuric acid (sulphur trioxide) instead of wholly into sulphurous anhydride. I base my opinion on the qualitative tests of the condensed matters, and upon the corrosive properties of the fumes rather than upon any profound chemical investigation of the subject. Inconclusive as my observations are, I find them much at variance with those of Mr. Austin, who speaks of the sublimation of elemental sulphur as an accompaniment of his work. I should be very much surprised to be shown the existence of unburned sulphur or any other oxidizable substance in the fumes of my pyritic furnaces, the phenomenon being entirely at variance with the conditions. I am so convinced of the copious formation of sulphuric acid in the pyritic furnace, at least under conditions familiar to me, that I somewhat expect to see the phenomenon made use of in the direct manufacture of that most valuable re-agent; for no great difficulties would appear to stand in the way of condensing the volatilized acid in cooling chambers or in towers, and its subsequent purification from the solid and liquid impurities which would naturally accompany it. *

40. FUEL ECONOMY OF SMELTING PROCESSES.—Regarding the relative economy of fuel by the various processes under discussion, it is evidently fair to assume *a priori* that that one which results in a complete combustion of the fuel will, other things being equal, surpass in economy those in which the combustion is incomplete. This assumption appears to be borne out by the experience had in pyritic smelting, where the fuel percentage is much diminished, irrespective even of the sulphides which may be burned.

41. DEDUCTIONS.—Upon the consideration that combustible matters generally are burned, including the volatile combustible products of the distillation of coal and wood, it follows that other

---

* It may be objected that sulphuric anhydride formed in the shaft would be broken up in volatilizing into dioxide, and oxygen. This may be true in regard to the major part; but that a considerable evolution of the stronger gas takes place is certain.

fuels than coke and charcoal may be used in this form of smelting. Even the most volatile gases may probably be made use of, such, for example, as ordinary illuminating gas, natural gas, etc. Following this train of reasoning I was led to experiment with wood, which eventually I used, not as a mere makeshift, but regularly and successfully in practical work. The innovation evoked the criticisms of certain metallurgical acquaintances, who, not understanding the drift of my work, and perhaps imagining that their own experience and knowledge covered the whole field, were disposed to sneer at what they considered a pitiable makeshift. The advantages in its use were threefold: first, obviating the waste by inferior oxidation, which advantage is inherent in pyritic smelting; second, the employment of a fuel cheaper per unit of calorific power; third, a useful mechanical effect in rendering the charge less dense and thereby facilitating the passage of the blast. In order not to mislead the reader I will add that I have not been able to replace more than half the coke with wood.* My experiments in the use of coal in the pyritic furnace have been but slight and inconclusive, nor am I aware of others of a more thorough nature; but from analogy there should be important economies in the use of proper kinds of coal, and no technical difficulties therein. My conclusion, however, was that a portion of the fuel should be of a strong dense kind, not burning too freely, but passing well down into the tuyere region, so as to burn there and produce the hearth temperature needed to perform the work of melting. The use of too light and fragile a fuel, which is consumed high up in the furnace, will in every case give rise to a cooling of the hearth and the formation of incrustations which defeat the intentions of the smelter. Then is required the application of greater heat to rectify the disturbance. We may get this by increasing the percentage of fuel, more particularly of coke. But the addition of more coke, while it increases the hearth temperature, decreases the amount of oxygen available for combination with the sulphides of the charge. Consequently more matte is formed, and with equal steps iron is withdrawn from the slag.

---

* See *Erdmann's Journal*, XVII., p. 471, for description of the practice at Nischui Tagilsk, where raw copper ores are smelted, and roasted matte re-smelted, by the use of wood in the blast furnace. Two hundred and fifty cubic feet are required to about 4½ tons of mixture. This instance, unknown to the writer until of late, clearly antedates his practice, while the objects are different in part.

48 MATTE SMELTING.

We are receding then from pyritic work and approaching plain
matting, wherein higher hearth temperature is obtained at the expense of low concentration of product. We have a better resource
against the emergency in the hot-air blast, which has been applied
to this form of smelting with great success by Mr. Austin, whose
older process depends mainly on the increased oxidizing effects
of hot instead of cold air.

42. Considerations Regarding Heated Blasts.—Hot blast
means less blast. For the heated air brings with it the heat which
would otherwise have to be generated by burning fuel at the
expense of a part of the oxygen of the blast; and the loss of this
oxygen diminishes the oxidizing power of the resulting atmosphere,
contaminated as it is with the products of combustion. From the
higher oxidizing power of the hot blast follow the use of higher
sulphur cont'nts of the charge; increased hearth temperature; lower
position of the zone of fusion; increased silica contents of the slag,
and hence lesc flux. A diminution of the oxidizable contents of the
charge brings about a diminution of the efficiency of the hot-blast
processes. For when the sulphides become scarce, the lack has to
oe made up by the addition of carbon, and we reach a point where
the process becomes virtually plain matting. Three kinds of fuel
are used at once in the Austin Oider Process: coal, wood or oil to
heat the blast, where the fuel cannot be burned so economically as
within the shaft; sulphides, whose fuel efficiency costs nothing
and which are better burned than not; and coke, to make up a
customary, or perhaps only occasional deficiency of heat in the
shaft. The heat from these three sources combined is sufficient to
carry on the reduction and fusion of a charge considerably more
refractory, it is reasonably claimed, than can be handled in the
ordinary blast furnace. This, f course, would be a very important
advantage in itself. The fuc for heating the blast is usually cheap,
and there is an additional but doubtful advantage claimed in the
unusual capacity of the furnace. Under ordinary working conditions this process has proved effective and cheap. Under more
favorable conditions it might and probably would prove one of the
cheapest in existence.

43. Temperature of Blast.—The foregoing considerations
relate mainly to those cases of pyritic smelting wherein the air for
the blast is heated by means of fuel burned in blast-heating con-
trivances outside of the furnace. The apparatus made use of in
such cases is similar in design and effect to the hot-blast stoves

used in connection with iron-smelting plants, but differing there-from in the character of the fuel used. For whereas, in iron production in blast furnaces, the gases evolved from the smelting contain a large proportion of combustible ingredients and are made use of as the fuel for the air-heating, we are in the pyritic processes debarred from this source of heat in consequence of the character of the products of combustion, which contain no substances susceptible of further oxidation. The practice has, therefore, been to employ ordinary fuels for this purpose, which are consumed upon grates beneath the cast-iron pipes of the blast stove. This is the course of procedure followed thus far by the pyritic plants on the Austin system. We have, however, a valuable and costless source of heat in the waste slags which flow constantly or intermittently from the furnace, which is entirely sufficient, if properly utilized, to heat the blast; and several plans have been proposed to this end, which will be adverted to in the proper connection.

I need hardly add that the methods of heating the blast are in no case distinctive of any process of pyritic smelting; and I may say further that the application of the hot blast to this form of smelting has not been patented to any person in the United States, and doubtless cannot be so patented. Metallurgists, therefore, are not debarred from the employment of the heated blast as a means of pyritic smelting, the only patented features of any process being solely those novel mechanical principles which distinguish the different processes.

Having provided ourselves with a costless source of heat for the blast, the limiting conditions of the discussion in part disappear, and the question of the availability of the hot blast becomes in the majority of instances a question of the cost of installing the apparatus. There are cases, however, where the hot blast from its intenser oxidizing power might leave no undecomposed sulphides for the matte; and here the cold blast would prove clearly the more serviceable. Accordingly, assuming that the blasts of low temperature comport best with quite small proportions of sulphides, it follows equally from other considerations that high sulphide content admits and even demands hotter blasts. Experience may yet establish the rule that the temperature of the blast should vary directly with the content of sulphides.

44. MECHANISM OF THE AUSTIN OLDER PROCESS.—As shown in the published drawings of the Patent Office the mechanism of the Austin Older Process presents some interesting peculiarities.

It consists principally of a shaft furnace of normal outward appearance, provided with a central tube inside the shaft, extending
downward from the feed floor perhaps three-fourths of the distance
to the tuyere level, and ending in the vicinity of the smelting
zone, affording a means by which the pyritous portion of the ore
may be fed directly into the smelting zone without having to
undergo the usual course of downward travel along with the remainder of the charge, and consequently without undergoing the
usual preheating which is an incident of the descent of the materials in other furnaces. The object in thus delivering the pyritous
material is twofold. First, to avoid the incipient fusion and consequent sticking to the walls which would ensue were the charge
to be fed as usual; and second, to prevent the partial oxidation
which would also ensue with deterioration of the fuel value of the
pyritous material. The annular space between the inner wall
proper of the stack and the tube is, or may be utilized to feed the
infusible portion of the charge, should it be required. It is to be
understood that the smoke and gases pass up and out of the furnace by way of this annular space, there being no particular
movement of air or gas within the central tube, although for convenience the function of the tube and annular space may be re-
versed and the gases allowed to pass out by way of the tube, while
the fusible constituents of the charge are fed into the furnace by
way of the annular space. Hot air is used for the blast, the heating apparatus consisting of a hot-blast iron-pipe stove or equivalent device. These modifications and additions to the customary
apparatus and process make the plant more costly beyond a doubt,
though it is denied by the patentee that it is more costly per unit of
smelting capacity. What I would more particularly call the
reader's attention to is the fact that the Austin Older Process
operates at the smelting zone on cold masses of base-metal sulphides, virtually burning them as lumps of coal are burned in the
blacksmith's forge, or in the boilers of a man-of-war under forced
draft. Under these circumstances a very high temperature can,
it is said, be attained, though this might appear to some inconsistent with the feeding of cold material directly into the zone of
fusion. It is consequently claimed for the process that difficultly
fusible slags may be successfully produced, which could not be
handled in any furnace blown with cold air. The advantages of
being able to make bi-silicate and still more acid slags have been
touched upon in another connection. Where an almost absolutely

clean slag is required it is undoubtedly a certain class of bi-silicates which most nearly meets the requirements, and in those numerous cases where silica preponderates there is clearly the greatest advantage in having at command a furnace and process capable of dealing with the acid slags resulting.

45. CONDITIONS OF WORKING. — If the reader will reflect upon the necessary condition of things in the smelting zone of this furnace he will be struck with its novelty. First he will perceive good reason why the smelting space cannot rise, producing fire-tops, insmuch as there is no fuel in the upper part to take fire, the only combustible materials of any sort being preserved from chemical action by being pent up in the interior tube away from the heat and the oxygen, until the time comes for their discharge into the fiery zone immediately below. They reach the active zone while as yet unchanged; and there is the best of reasons why a gradual oxidation of the combustible would not answer in this system of smelting. The sulphides might indeed be hot when at the moment of feeding into the smelting space with good advantage in one respect; but this would presuppose a partial decomposition, with escape of sulphur and oxidation of metals and consequent loss of calorific power, which Mr. Austin seeks to avoid. Again, we need no hint from the inventor to enable us to discover that an ore consisting of sulphides disseminated in a stony gangue would be ill suited for this style of furnace. For the requisite rapidity of combustion could not take place with an ore whose combustible particles are hidden and protected by non-conducting material of this nature. Nor need we seek further for the reason why fine-grained substances are not successfully treated. We are entitled to assume that only solid, coherent masses of sulphides or arsenides, undiluted by gangue, especially gangue of an intractable, stony sort, are calculated to work with the best success in the Austin furnace, and that fine ores and disseminated sulphides, no matter even if the latter are chemically unexceptionable, will not serve the requirements of the process. It would also seem that a reasonably fine comminution of the remaining part of the charge, that portion in fact which comes down the annular ring and through which the waste gases pass, heating it and preparing it for combining with the other components, might be a very good precaution, considering how short tho time for complete reduction and combination is, and the highly siliceous character of the ore which it is an especial purpose of

this furnace to slag. We may venture another assumption, namely, that the success of the operation depends very largely upon the separation of the components of the charge into two parts, the one of combustible matter, of sulphides without gangue, the other of gangue without sulphides; and that as these become mixed, or are naturally mixed, the operation is less successful. As the inventor has published little about the theoretical principles which underlie his ingenious process, we are entitled to these assumptions and to as full and free a criticism as we may wish to pass. We are not entitled, however, to assume that, as charges of a certain description are likely to give the best results, nothing can be done with less favorable ores. On the contrary mixtures of quite a decidedly unfavorable appearance have been treated, hitherto impossible of profitable utilization, among others the sulphides of a Leadville mine, which are described as "a kind of blue mud." Mineralogically this description is doubtless inexact; but witness the proficiency of American metallurgists, in whose hands blue mud becomes at once a valuable ore and an esteemed fuel!

46. LATER MODIFICATIONS.—It is to be observed that the results achieved of late by Mr. Austin have not been in line with his earlier efforts. He has recently discarded the inner feeding tube, and in lieu thereof resorts to feeding the fusible materials into the center of the stack, while the infusible substances are placed around the periphery so as to separate the sulphides from the walls and so prevent them from sticking on when they reach a state of partial fusion. The supposition evidently is that the particles maintain their relative position during the downward passage, mixing together finally after arriving in the zone of fusion. One may be skeptical as to the value of the expedient for preventing the materials from attaching themselves to the walls, even when a considerable portion of the charge is composed of infusible matters; but we may disregard the query as unimportant. It is more pertinent to inquire what is to be done when the charge is largely or entirely made up of sulphides. For in such not unprecedented cases there would be nothing to prevent the fusible particles from coming in direct contact with the walls, when the distinctive features of the Austin process would disappear and the operation would become identical with those which I describe as the Gradual Reduction processes. I have accordingly classified Mr. Austin's newer process with them. Having discarded the central tube, the

only novelty lies in the patented method of feeding as before described, which is not sufficient to constitute a different system of smelting, and indeed does not vary practically from other methods now known and used.

47. MODES OF FEEDING FURNACES.—I advocate and practice the method of feeding by layers, as in lead and copper smelting, while I attach little importance to exactness in placing the materials so long as they are in position to become mingled during the fusion. Mr. Austin adopts such a method of feeding as will in his opinion prevent the commingling of the materials and the contact of the sulphides with the walls. Should incrustations form, which are inevitable under most circumstances, I find no difficulty in removing them by mechanical means, the furnaces which I employ being of slight depth and easily accessible from the feed floor through all their interior.

It is apparent that no great differences exist between methods of smelting described, whence it follows that there can be no great differences in the effects produced. By discarding the feeding tube the Austin new method loses wholly the peculiarity of feeding cold unchanged sulphides directly into the zone of fusion, substituting therefor the gradual heating and oxidizing effects common to other methods, and loses likewise in part the other peculiarity of retaining the sulphides in a central column.

48. COMPARATIVE TABULATION OF EFFECTS.—

AUSTIN OLDER PROCESS.	GRADUAL REDUCTION PROCESSES.
Characterized by—	Characterized by—
Sudden oxidation.	Gradual oxidation.
Contraction of smelting zone.'	Expansion of smelting zone.
Concentration of heating effects.	Diffusion of heating effects.
Rapid transference of heat.	Slower transference of heat.
Evolution of sulphur and sulphurous anhydride. (?)	Evolution of sulphurous and sulphuric anhydrides.
Favors hard blowing and rapid driving.	Favor lighter blast and slower running.

49. PYRITIC SMELTING OF SIMPLE ORES. — In its simplest form pyritic smelting consists in charging a mixture of quartz and pyrite, and oxidizing so much of the latter that the resulting oxide of iron will suffice to take up the quartz, forming a slag, while the unoxidized pyrite is reduced by it to ferrous sulphide by the volatilization of one equivalent of its sulphur, and separates as matte, along with the gold and silver. It was said to have been in this

form that the Austin process was first worked out at Toston, Montana, where it was carried on subsequently in a commercial way. We are informed by the inventor that the charge which best brings out its advantages contains about twenty-five per cent. of sulphur and in the neighborhood of fifty per cent. of siliceous vein matter, which furnishes silica enough to make a bi-silicate slag, which he regards as possessing peculiar advantages. To deal with a mixture of this description a hot blast is justly considered indispensable. For while we could, no doubt, increase our fuel ratio and the height of our furnace to enable us to handle a bi-silicate of iron slag, under these conditions the slag would surpass the bi-silicates in acidity—in fact, we would get little or no base formed, the oxidation of sulphides ceasing, and hence practically no slag could be formed at all. There has also been recommended a mixture of even greater acidity, to wit: equal parts of quartz and iron pyrites. Now, I imagine that this charge (which would give rise to a quadri-silicate of iron, if all the iron were to become oxidized in the operation) would prove impracticable even with the very hot blast which the inventor prescribes.

Recapitulating slightly, we can distinguish two different uses of the hot blast: first, for its sensible heating effect; and second, for its chemical effect aside from heating. The pyritic smelter's use of it is primarily for its chemical effect in burning the sulphides, which it performs efficiently. The sum total of the different heating effects goes to fuse and slag some of the solid constituents and to matte the others. This is pyritic smelting in its extremest development, to carry out which to the fullest extent requires a charge composed of coarse pieces of raw pyrites and quartz, the sulphur contents preferably reaching twenty-five per cent., or over. A highly heated blast must be employed. Under such circumstances it has been found possible, writes Mr. Austin, to run continuously for a time without using any ordinary fuel in the furnace shaft. Under less favorable, and as I understand it, ordinary circumstances, the fusion cannot be effected without the regular addition of fuel— say two or three per cent. of coke—to the charge.*

50. RELATION OF BLAST TEMPERATURES TO FUEL. — As we

* The writer would not like to be quoted as affirming or denying the allegations of those who have exploited the Austin processes, but he is abundantly satisfied that their claims to work at times without the use of any carbonaceous fuel in the furnace are well founded; as no doubt exists that the blast, if

decrease the amount of sulphides in the mixture, a greater and greater addition of fuel becomes necessary, until we reach a point where the sulphides become too small in quantity to have much effect as fuel, and indeed have to be preserved unoxidized to form matte. In this exigency the hot blast ceases to perform the particular chemical function allotted to it and becomes useful for its direct heating power alone. But as fuel expended in heating the blast outside the furnace can never in the nature of things be so economically expended as in heating it inside, it becomes evident that the hot blast may be no longer economical, unless indeed the fuel used in heating it in the hot-blast stove be very considerably cheaper per unit of calorific power than that which we are accustomed to use in the shaft. I do not present this last consideration as of great weight since, as previously indicated, I have found by experiment that some of even the cheapest and crudest fuels may under proper conditions be profitably employed in the pyritic furnace, taking the place, at least to some extent, of the more costly artificial fuels in common use. While, therefore, no one can deny the utility in appropriate circumstances of the hot blast, it is not to be indiscriminately recommended, in view of the character of much of the ores, which makes its application superfluous.

51. ADAPTATION OF PROCESSES.—Mr. Austin's idea of the proper functions of pyritic smelting might be esteemed somewhat radical by those familiar with existing conditions in the mining regions, for his intent has apparently been to devise a means of treating ores or ore mixtures carrying a very high proportion of sulphur, much higher in fact than are found to occur as a general rule. The experience is that base metal ores of so high a tenor in sulphur are rather the exception in our Western mining districts, and that the bulk of the so-called base or refractory ores of gold, silver, and copper will not exceed 15 per cent. sulphur, and possibly will not average 10 per cent. It is probable that a metallurgical plan which is adjusted to the treatment of such moderately pyritous material will prove more generally advantageous than any attempts to meet extraordinary conditions by extraordinary means.

---

heated enough, can take the place of such fuel and produce all the effects desired ; but that it is feasible to treat the average run of base metal sulphides without fuel is more than doubtful. In fact the language now used by Mr. Austin in reference to his invention would-indicate that he has abandoned his former claims in this regard and now makes no pretense to smelt without fuel.

52. PREVAILING CHARACTER OF SULPHIDE ORES. — It is seldom that we find very large quantities of pure sulphides of any sort; pure pyrite, pure chalcopyrite and pure blende rarely exist in nature uncontaminated by other substances, but are found intermingled in the most heterogeneous way with each other and with the other sulphides, with arsenides and with the various gangue matters. The prevailing characters of base metal ores which seek beneficiation are mixtures, most commonly silver-but occasionally gold-bearing, with or without copper, which exists often in small, but not unfrequently in important proportions, and which is saved with the same facility, of course, as the precious metals. The very important subject of gangue merits attention. We find a large proportion on the average of stony matter, perhaps not less than two-thirds, in those ores under consideration, and we classify those of a still lower content in base metals as concentrating ores, as being best adapted to water treatment. The gangue is prevailingly quartz, but in more favored localities is calcite, spathic iron (seldom), heavy spar (less seldom), and not unfrequently is some felspathic rock, usually somewhat decomposed. Here is indeed a formidable list of substances, difficult enough formerly to deserve the opprobrious epithets rebellious, refractory, base, etc., which our metallurgical predecessors with good reason showered upon them. But while difficult and intractable enough when taken singly, or when dealt with by inefficient means, their difficulties vanish when assailed by the overpowering forces of the pyritic furnace, which conquer and dispel every opposing element and combination. The rebellious qualities, or what were deemed such, of the ore are the foundation of pyritic smelting, which utilizes hitherto objectionable elements in its essential reactions, making use of the very characteristics which make all other processes inoperative or highly expensive, and rendering those substances which by their nature are the most difficult of beneficiation by other methods, the very cheapest and best adapted to the pyritic treatment. It is, or will be, the rule that those ores which are the most intractable to other processes are the easiest to treat by pyritic smelting.

53. USES OF THE COLD BLAST. — Cold blast methods of pyritic smelting, which I put forward not as rivals of, nor substitutes for the hot-blast methods, furnish a rational and efficient means of dealing with those classes of base metal ores which do not contain the highest percentages either of sulphur or silica, but which are so rich in the former that they are not susceptible of

direct treatment by other processes. An ore mixture containing 10 per cent. of sulphur, combined mainly with iron, might in the presence of sufficient metallic bases be expected to produce, roughly speaking, about 30 per cent. of matte, if run down in a furnace with strong reducing action. This would be the customary result of the treatment of such a charge by the German system. The 10 per cent. mixture, however, may be fused in a furnace in accordance with the principles of what I have called the cold-blast pyritic system, and the amount of matte reduced to 10, 8, or even a less percentage, the work being attended with the oxidation of a corresponding proportion of metals and the volatilization of sulphur. Thus, while the German system, applied to certain pyritic ores, effects a concentration of three into one, or thereabouts, the cold-blast method of pyritic work puts ten into one, or twelve into one—a result that necessitates burning off three-fourths or more of the sulphur and the oxidation of nearly half the iron. That such results can be regularly and systematically effected without the agencies of the hot blast, but simply by changes in furnace construction and manipulation may seem incredible at first glance, but practical experience amply confirms the assertion. I was first led to experiment in this direction by noticing the variations in the amount of matte produced by furnaces at different times, which variations I traced to the influence of the incrustations formed in the hearth and upon the walls. The amount of matte produced is to that extent an index of the chemical changes, and hence of the condition of the atmosphere within the furnace. Recognizing the relation between the form of the furnace shaft and the oxidizing or reducing effect, I was enabled to apply those modifications and improvements in process and apparatus which constitute what I have called, I hope justifiably, a praticable cold-blast pyritic process.

54. PRINCIPLES UNDERLYING COLD BLAST SMELTING.—This form of smelting depends upon the circulation of free oxygen through the charge, making it necessarry to so apply the blast that such circulation is assured. For this purpose I admit the air into the blast furnace in a particular manner and in various quantities, as may be required for the particular kind of ore in hand, the amount of sulphur being the principal determining factor. Having this active circulation of air, especially in the upper part of the charge, it becomes possible to effect a striking degree of oxidation and consequently of concentration. The best results that I

have thus far secured in practice are when treating charges containing a medium amount of sulphur, say from 8 to 15 per cent.,
although there is no necessary reason why still higher percentages
should not be successfully dealt with.   While eminently successful
on ore of this medium description, the lack of higher sulphureted
kinds has thus far prevented the demonstration of what the cold-
blast apparatus can do with those excessively base compounds which
are preferred for the hot-blast methods.   A further improvement
recently projected and not yet put in practice, will, I am confident,
furnish a powerful means of reduction, suitable to either the hot
or cold-blast processes.*   In the most successful work the slag was
of a fusible sort, the bases being largely protoxides of iron and
manganese, for average analyses of which see the accompanying
table.   These slags ran freely, the charge requiring 7 to 9 per
cent. of fuel for complete reduction.   As shown, the slag contained
on the average one and a half ounces of silver per ton—not as clean
as many that can be made, but the most economical possible under
the conditions.   Whereas Mr. Austin finds it possible to produce
bi-silicate slags whose base is almost exclusively iron, I found that
slightly acid silicates of the same base (and manganese), or
sesqui-silicates of mixed bases best satisfied the conditions, excepting probably as to the freedom from gold and silver, in which
respect information is somewhat lacking.   By employing these
means we may, without the help of the heated blast, free the
smelting mixture from much the larger part of'the sulphur, and
oxidize an equivalent proportion of the bases; but this must be
done under conditions which, so far as I can see, preclude the employment of more than moderately refractory slags.   It becomes
necessary to maintain such a composition that the slags are not
greatly less fusible than the ordinary lead slags.   On the other
hand we can treat fusible basic mixtures with great facility.

---

* This is the writer's invention of concentration of matte by returning it
while molten to the blast furnace.   It is adapted to blasts of any temperature.
Poured upon the charge in the shallow furnace it comes immediately into contact
with active gases which decompose it partially, with evolution of sulphur
gases, and production of oxides.   This is mainly practicable where the charge
is shallow, and hot nearly to the surface.   The furnace atmosphere must
necessarily be oxidizing in its effects.   Besides the concentration of the matte,
there are secondary results of value, first in supplying a portion of matte to
desilverize the slags in the event of the momentary cessation of matte formation ; and second, a useful effect upon the furnace hearth whenever obstructions tend to collect.

55. DEMONSTRATION OF PRINCIPLES.—I will endeavor to explain the principle which governs the case. The function of the boshes in a furnace is well understood. They are to concentrate the heat and the chemical action within a small space, whereby intensity of combustion and reduction result. The air from the tuyeres is brought into contact with the fuel and materials of the charge, whereby the oxygen is entirely consumed, with great intensity of heat but with imperfect combustion of the fuel. A matting furnace having a contracted zone of fusion volatilizes little or no sulphur, but reduces oxides powerfully and makes much matte; and, I may add, is able to produce very hot slag. We may infer from this that to remove the boshes so as to enlarge the zone of fusion horizontally would have the effect of diminishing the reduction of oxides, and of volatilizing sulphur, and consequently diminishing the production of matte. This is, in fact, what does take place; for the oxygen of the air, instead of being forced into contact with white-hot carbon and thus consumed, finds spaces in the enlarged zone of chemical activity through which it can escape upward, whereby it is enabled to combine with sulphur and other components of the charge for which it is avid, and so carry on those reactions which are the distinction of the pyritic process. But notice that the enlargement of the zone has also the effect of reducing temperature, bringing in the evil of cooler slag, and consequently debarring us from the employment of difficult mixtures.

56. FURTHER DEMONSTRATION OF PRINCIPLES. — The most rapid and efficient reduction of ore in a blast furnace takes place after the smelting zone has been so narrowed by the formation of incrustations upon the wall and about the tuyeres (noses) that the air, first warmed by passing through the slag which has collected about the inlets, is forced into contact with the white-hot fuel, when the latter burns with powerful intensity of action, creating a great heat within a comparatively small space. In a round furnace smelting a hundredweight of materials in a single minute the melting may all be performed within a space the size of a flour barrel. Within a certain limit a narrowing of the active space is accompanied by a rise of the temperature; and a horizontal increase of the smelting area is followed by a decrease of the temperature. But such an increase or decrease of the horizontal section of the active smelting space is also accompanied by a corresponding though inverse change in the production of matte. Now,

matte production we know varies inversely as the oxidizing action
of the furnace upon the sulphur and other matte-forming substances in the charge. We may also note in this connection the
dependent consequences of the addition to, and withdrawal from
the slag of the bases which result from the varying oxidizing effects
described. I noticed that a certain matting furnace, when freshly
blown in and therefore perfectly free from crusts, yielded very
little matte; while after the accumulation of crusts began the proportion of matte rose rapidly, in strict accord with the contraction
of the smelting area, until when the latter had been reduced to
about one-third of the space it at first occupied, the output of matte
had grown fourfold at least. The temperature of the slag, which
at first was lower than favored the requirements of good work, rose
also, while it increased perceptibly in acidity, owing to the ab-
straction of iron oxide. It follows from these considerations that
oxidation can be furthered in blast furnaces by simply retaining
the normal smelting area, by preventing or removing the wall
incrustations which have so strong a tendency to form.

57. FURNACE CONSTRUCTION AND MANAGEMENT.—In designing furnaces for cold-blast pyritic work the metallurgist will of
course eschew the bosh, and with it the downward tapering circu-
lar stack of the Arizona copper furnace type, as being possessed
of qualities opposed to those desired. For this form he will sub-
stitute a prismatic shaft whose width is such as to allow the blast
to penetrate all parts; whose length is proportioned to the quantity
of work desired; and whose depth, being governed by the fact
that fusible slags are to be made, need be but slight. The sides
are vertical plane surfaces upon which obstructions do not readily
collect, or having collected, may be easily removed by means of
tools worked from the feed floor above. These simple means
serve perfectly the purpose intended, and the principal requirement of this kind of smelting, namely, the preservation of
the original interior form of the furnace, is accomplished, whereby
the work is kept under control, the production of matte remains
normal, and the results of the operation, thus rendered certain,
respond in all respects to anticipation. To the experienced metallurgist I need hardly remark that a process whose results cannot
be closely foreseen as we pursue it step by step is not apt to prove
of any great value in the arts. I may say of the pyritic processes
that their value and availability depend upon the skill of
the attendants and the watchfulness of the persons in charge to

a higher degree than any others with which I am acquainted. I cannot commend those processes as being easy to conduct. Chemical analysis, which gives so perfect a control over the lead and iron smelting, is a far less reliable guide in pyritic smelting, and is far from enabling us to predict the results of contemplated operations. In the absence of actual experience (of which nothing can take the place) calculations based on analysis are in a measure untrustworthy. At the same time I would not be understood as proposing to dispense with chemical work, which is from my point of view indispensable. To bring out the full advantages of the processes, there is required an extensive experience supplemented by watchful care and the assistance of chemical analysis. Without these failure, or at least a long and costly term of experimentation, is inevitable. Considering the small number of metallurgists who have thus far had actual experience in pyritic work it will be safe to prophesy occasional failures, much dissatisfaction, and loss of time before a sure foundation for success is reached.

58. THE FURNACE BLOWER.—Returning again to the subject of furnace construction, I have to remark that to attain efficiency in pyritic smelting the requirements of the blowing apparatus have to be modified to correspond with the furnace. The depth of the charge being but slight there is no need of the powerful apparatus to which we are accustomed in other branches of smelting. We need, however, a very great volume of air; and this, at the low pressure necessary, is best supplied by the Sturtevant blower, or some similar centrifugal fan. With much fine ore it might be advisable to use a very large rotary blower instead. Throughout the designer will have borne in mind especially the proportion of oxidizable materials in the proposed mixture and provided such means for supplying the air for oxidation as will be necessary in the given case. Should the sulphur exceed the proportion supposed to be manageable by ordinary means he has the choice of several methods for its removal. Mr. Bartlett and myself have practiced the introduction of an auxiliary blast at a suitable level, to be determined by, and adjusted in accordance with the condition of the interior of the stack. One may, probably, also very expeditiously concentrate the matte by returning it while still liquid to the charge, it being then in the best condition for burning in the blast from the tuyeres. This conception, previously adverted to, seems to possess great advantages from its sim-

plicity and probable effectiveness. It, however, has not got
beyond the incipient stage as yet.

59. COMPARATIVE RATE OF SMELTING. — It has been held by
some that the smelting capacity of pyritic furnaces is greater than
that of ordinary furnaces of equal size of hearth. The writer
must dissent from such views, which are contrary to experience
and reason. It may be that furnaces intended to work on the
pyritic plan have been driven at a faster rate than other furnaces
have been on the same description of charge. But this degree of
speed is inconsistent with the attainment of those concentration
effects which are our aim. In fact the work could hardly be
called pyritic smelting at all, but would fall into the category of
the German system, wherein no oxidation of the ore is expected.
Time is an important element of pyritic work, and there exists a
direct relation between the oxidizing effects produced and the
duration of exposure of the ore to the air currents, and on the
other hand an inverse relation of both to the amount of matte
formed. All the experience yet had, so far as I am aware, confirms this view, and I have no hesitation in saying that pyritic
smelting (excepting possibly the Austin older process, of which I
do not here speak) is necessarily a slower method than the German
system of smelting.

This fact, which to some might appear to injure the efficacy of
the system, is not in reality a vital nor even an important objec-
tion. Granted that it be slower even in the proportion of one to
two, it is only necessary then to double the hearth area in order to
restore the smelting capacity. Nothing else about the works need
be changed; blower, power plant, working force, all remain at
the normal.

60. HEARTH ACTIVITY.—We must remember, however, that as
there is a limit beyond which a furnace cannot be driven, there is
also one below which it cannot fall without injury to the smelting
process. The diminished rate of smelting which favors oxidation
and high concentration ultimately reaches a point where no smelting at all can take place and the process comes to an end. The
minimum amount which can be treated in a blast furnace in a given
length of time, still keeping the slag and hearth hot enough to
work acceptably, is an interesting question which we cannot afford
to leave undiscussed. We can compare the smelting powers of
furnaces as to their total capacity in units of charge, or, better and
more scientifically, as to their capacity in units per square foot of

tuyere section. We may designate this intensity per square foot
as hearth activity; and I obtain its value in each particular case by
dividing the consumption in tons of all solid materials which go
into the furnace in twenty-four hours by the tuyere area of the
furnace. Thus I find the hearth activity of the Sudbury furnaces
to be about 7; that of a matting furnace at Mansfeld, 5; at Mineral, 3 to 4; Orford, 2¾; Bisbee, Arizona (copper), 7; while the
hearth activity of the Edgar Thomson furnace "I," making iron,
reached 14. The lowest feasible rate is probably about one ton per
foot, of which the old European stacks furnish several examples.
No modern furnace either here or abroad runs so slowly, nor can
good clean smelting be done at such a rate. Probably the needs
of the pyritic system would be best subserved by a rate of driving
of from two to four tons per foot, according to the condition of
the charge, etc. Since it is an indispensable requirement in this
style of work to drive slowly, which has a tendency to cool the
hearth, I have suggested as a means of keeping up the temperature
of that part of the furnace the return of the molten matte according to my heretofore described invention. By this means the very
slowest rate of smelting could probably be achieved whenever
deemed desirable, and the hearth kept sufficiently hot for the purpose by the repeated passage of the matte.

61. PRODUCTION OF PYRITIC EFFECTS.—When we use an air
blast of large volume and low pressure, especially if it be cold, the
conditions are favorable for the formation of slag crusts about the
tuyeres, and a partial filling up of the hearth with the hardened
masses, which are perforated with holes through which the air
currents pass upward. In such a case the tuyeres usually show no
light. The conditions are now favorable for pyritic work, as long
as the fuel is kept at a minimum. They are also good for chilling
up, which has to be guarded against, not as being a serious calam-
ity, however, which it is not, for properly constructed furnaces are
easily and quickly relieved of a chilled charge. The ease with
which the crucible can be removed, the charge dropped, incrustations knocked off, a hot spare crucible placed in position, some
fuel given, followed by a few hundredweight of slag and matte,
and then the ordinary charge, and operations resumed by letting
on the wind, is such that it becomes cheaper sometimes to blow
out than to bar down, although the barring is proportionally
easy.

62. SECONDARY EFFECTS.—The large volume of cold air at low

pressure produces secondary effects which have an important bearing on furnace construction. A good proportion of the blast fails
to enter the charge to any distance, but finds its way to the surface
by following up the walls where the charge lies loosest, and producing a cooling effect there which prevents fusion, while the
excess of oxygen serves to effectually roast a proportion of the ore.
In the next inward layer the air is less abundant, a higher heat
prevails, and smelting takes place combined with oxidation of the
combustibles of the charge. Further inward the tendency to combustion becomes still less, owing to the scarcity of air, and the
roasting and smelting effects both cease, the former first. Were
the furnace wide enough, a core of inert matter, unacted upon by
the blast, would exist in the center, and to avoid that result such
furnaces are made narrow. It will be evident that the most
thorough oxidation will take place in contact with the walls,
or a little way in. Therefore if we desire the highest concentra
tion we must increase the extent of the wall area by lengthening
the furnace, while we diminish the breadth at the tuyeres to get
rid of the inert space. This is the principle of peripheral extent,
which cuts an important figure in pyritic work. Mr. Bartlett has
shown his appreciation of its value by designing a furnace having
an interior length of sixteen feet, with a breadth at the tuyeres
of two.

It also follows from what I have said about the effects of the
copious air blast, that water jackets are not necessary in all cases
to protect the furnace wall. The cooling influences of the blast
causing the deposition of slag incrustations on the surfaces most
exposed to danger, and protecting the walls above by cooling below
the fusing point the ore which rests against them, there is no
reason for the employment of other than brickwork constructions.
It is my firm conviction that water jackets are used in many
situations where brick or stone work would answer better; and
that their use is a fashion which is carried to excess, in an age
when the properties and usefulness of fire-resisting materials are
so thoroughly understood.

63. THE CRUCIBLE AND THE FOREHEARTH. — The forehearth,
which plays an important part in the German system of smelting,
is in some respects superfluous in pyritic work. In performing the
separation of matte from slag outside the furnace the thing aimed
at is to avoid the deposition of "sows" or "salamanders" of
metallic iron in the crucible. But as there is no tendency toward

the precipitation of any metallic substance whatever in pyritic work, we have reason to discard the forehearth and return to the use of the interior crucible, which answers the purpose better in most respects. We lose less heat by radiation, and the apparatus is easier for the men to handle. It is conceded also that the separation is better, as it certainly should be, as it has the advantage of a higher temperature. I have not been able by experiment to satisfy myself that it really is better, however. I much prefer the movable form of crucible, which admits of being changed about with as much facility as the forehearth, whenever reason exists, but as a rule it does not require it so often.

There should be a tap-hole each for slag and matte, in the end or side of the crucible, and the matte tap should be at least twelve, and better fifteen inches below the slag tap-hole. With a less distance it is difficult to obtain pure material at each tapping, as the vortex caused by the rapid flow of matte drags the overlying slag out with it, even when abundance of matte remains in the crucible. The aim at a well-conducted matting plant should be to make two products: pure matte and pure slag; and the proper management of the products as they issue from the furnace is a subject that will repay hard study. The slipshod practices at some lead smelters, and also, I am sorry to say, 'at some matte smelters, where immense quantities of mixed slag and matte are undergoing sorting, and unlimited amounts of foul slag are forever on their way to be re-smelted, should have no place in modern practice. "Sorting" and "cinder picking" are only necessary when the furnace work is badly done, or when the plant is ill-arranged. The re-treatment of slags can ordinarily be confined to that small proportion which, by reason of its favorable action on the furnace, or from its accidental richness in valuable metals, it becomes advisable to re-smelt. But the wholesale re-smelting of slags in order to recover mechanically mixed matte is a practice far behind the age.

64. BLASTS OF HIGHER TEMPERATURE.— The effect of the heated blast upon charges high in sulphureted constituents is remarkable and quite beyond anticipation. Even air blasts only warm to the touch produce effects so much more intense than cold ones that the difference is almost unaccountable. It has long been remarked that from the heat of summer to the cold of winter, blast furnaces engaged, for example, on lead smelting, fall off much in their duty per unit of fuel; and the difference in the pyritic furnace is much more striking still. In fact, the possible

difference of a hundred degrees Fahrenheit in blast temperature is
of great significance in this kind of smelting, where the object is
to secure the greatest possible production of heat at the lowest
attainable level in the furnace, and with the least contamination
of the internal atmosphere by inert gases. Not only the temperature but the purity of the air blown in exercises an important influence. The active agent being the oxygen of the air, it is
evident that deterioration in this regard will be followed by injurious consequences, and those inventors who have proposed to
use the partially deoxidized and exhausted air which has once
supported combustion will here find themselves much at fault.
There is exhibited a very inadequate understanding of the means
which are required and the changes which are wrought in a smelting
furnace, when it is gravely proposed to heat the blast by means of
a jet of oil burning in the delivery pipe at the expense of the oxygen of the air which is being blown; and equally so to attempt to
derive a hot blast by blowing in the exhausted and vitiated gases
from another furnace, worthless and ineffective as the expedient
must be.

Touching the increased oxidizing power of the heated blast, we
may discuss some statements taken from Kerl showing the effects,
relatively considered, of blasts in use at two foreign works where
the matting is of the old or German blast-furnace description,
(German system). Thus, at Mansfeld it was found that the
heated blast of highest temperature favored a more complete
oxidation of the carbon, producing in consequence a greater proportion of carbonic acid and less carbonic oxide than the cold
blast. The gases from the cupolas of Reicheldorf, where a warm
but not hot blast was in use, contained about 25 per cent. of combustible substances and a total calorific efficiency of 58 per cent.
was attained, while the blast of unheated air gave but 50 per cent.*
Such a result as the latter should have afforded every encourage-
ment to the inquiring metallurgist to proceed further on this line,
reducing the fuel ratio and increasing the blast temperature until
the experimental results had developed something of importance.
Pyritic smelting lay dormant in this ground, but Reicheldorf was
not destined to be the scene of its discovery.

65. METHODS OF HEATING THE BLAST.—The obvious advantages of the heated blast as applied to various forms of cupola

---

*Kerl, C. & R. translation, *Copper*, p. 197.

smelting have led to many new suggestions for, and improvements
in its production. Several schemes have been devised for securing
a hot blast by intercepting the waste heat from the walls of the
furnace, the upper part of the stack, the crucible, or the forehearth, as the case may be. Some of these are obviously inoperative from complexity of parts; while others are so from the fact
that the heat radiated from the apparatus is too small in amount
to be practically useful, even if all of it were absorbed by the air
blast. The forehearth and crucible of a blast furnace may, and
often do radiate enough heat to make the bystanders uncomfortable;
but all of the heat so lost to the process is hardly enough to make
it worth while to take measures for its recovery. Again, it has
been proposed to subject air in pipes to the influence of the heat
contained in the melted materials in the forehearth, with the hope
of thus providing a heated blast. But we must not forget that
the loss of heat by the mingled matte and slag would be fatal to a
proper separation of the two. We dare not abstract heat from the
forehearth, which in ninety-nine cases in a hundred is too cold
rather than too hot, notwithstanding all the nursing which we can
give it; nor from the crucible, which is hardly more likely to
afford the requisite supply of heat. This being the case, it has
seemed to me that the only available source lay in the slags after
their discharge from the furnace or forehearth into the pots, where
they are exposed to cooling by radiation. Here is a supply of
heat, which, could it be thoroughly utilized and returned to the
furnace, would enable smelting to be carried on without any fuel
whatever other than the due proportion of combustible ore. My
device for heating the blast by means of the waste heat of the slag
consists in an arched heating chamber of considerable length built
between the furnace and the slag dump, and so located that the
pots as they are filled are enabled to pass longitudinally through
the chamber, losing their heat as they proceed. The chamber
contains in its lower part a railway track on which the slag cars
run, and in its upper an air main, both being parallel to the axis
of the chamber. This arrangement would be inoperative were it
not that it is divided into successive compartments by vertical
partitions so as to be heated to different degrees, that compartment
nearest the furnace being hottest, whereby the air in the main is
subjected to an increasing temperature and the slag is enabled
to cool effectually before it finally leaves the chamber. The rate

of travel of the cars and the length of the chamber determine what amount of heat is communicated to the air in the main.

66. COMPARISON OF THE HOT AND COLD BLASTS.—Briefly comparing the advantages and disadvantages of the hot and cold blasts, I should say that there appear to be two situations wherein the hot blast is indispensable in pyritic work. The first is when the slag is of a difficultly fusible sort; the second when the sulphur contents are very high. In either case the hot blast will undoubtedly confer benefits far beyond the cost and complexity of its installation. There are likewise two cases in which the cold blast may perform its work best. The first of these is where the sulphur contents are low, say not above 8 per cent, or where the oxidizable constituents are so small in quantity that they have mainly to be preserved unburned to form matte. Second, in the absence of experiments bearing directly upon the matter, I can only suggest as something probable that the cold air blast will admit of the saving of a larger proportion of lead than the heated blast. I take it for granted that it is so, and that it is a matter of some significance.

67. DIRECTION OF EXPERIMENT.—It may be of interest to the reader to learn under what conditions of pressure, temperature, etc., the various successful forms of pyritic smelting were conceived and worked out. Thus, Mr. Austin's former process was carried on by the use of a very hot blast (perhaps 1,000 degrees F.), of moderate or small volume, and rather high pressure. His present tendency appears to be to lower temperature of blast and the regular use of inside fuel. Mr. Bartlett, who formerly used a high pressure and moderate volume of cold or slightly heated air, has, as evinced by his adoption of furnaces of great peripheral area, increased the volume of blast. His practice has tended to the use of two rows of tuyeres—a characteristic which has evoked much criticism, but which is founded in reason and experience. Those who have witnessed the performances of such an apparatus are amazed at its powers of oxidation. My own work was principally done under these conditions: a cold blast of moderate pressure (generally eight ounces) and large volume. Enlarged experience and opportunities lead me to prefer high temperature, large volume, and low pressure; although as to the question of temperature in particular, it would have to be settled by practical considerations quite outside of the desired furnace effects. For example: an exalted temperature of blast attained

by the use of outside fuel, however advantageous its effects might
be, would very probably prove less economical than a more moderate
temperature produced by the waste heat of slag or by other costless means.

68. LEAD IN THE PYRITIC FURNACE.—We have seen that in
the German system and in lead smelting, the lead is saved, in the
first as a constituent of matte, in the second as metallic lead, in
consequence of the strong reducing action of the furnaces. I need
say no more upon the subject of lead in the pyritic furnace than
to remark upon the absence of those reducing agencies and to call
the reader's attention again to the presence of those conditions
which ensure the oxidation of so many substances. The conditions
are, as might *a priori* be foretold, adverse to the recovery of lead,
as that metal is scorified and driven into the slag, or even, under
certain circumstances, volatilized in an oxidized form, and this
in proportion to the relative oxidizing power of the apparatus and
blast. I would like to be thoroughly understood upon the latter
point. A very high degree of oxidation, assisted, for example,
by a blast of high temperature, produces from lead-bearing ores a
matte carrying but a small part of the contained metal. A blast
of lower temperature, involving the use of inside fuel, and consequently producing a less intense oxidation, allows a higher percentage of lead to go into the matte. In general, the loss is owing to
the formation of oxide of lead, which necessarily becomes the
silicate, and mingles with the other silicates as slag. Mr. Bartlett,
by means of special appliances, which secure a more intense oxidation, has carried the operation a step farther in this direction,
and has succeeded in volatilizing his lead (and I may add, his zinc
also) in a form which renders the sublimate of commercial value.
I again call the reader's attention to the description of this
interesting process which appeared in the *Engineering and Mining
Journal* as previously cited. Mr. Bartlett's work as described
therein is not strictly pyritic smelting—it is more than that. It
is pyritic smelting combined with the intentional sublimation of
a large proportion of the valuable ingredients of the charge, and is
practically the combination of two processes—the volatilization
of the lead and zinc, and the matting of the copper (and silver
and gold if present)—in one operation. I will not enter upon
any observations in this ingenious, and, I believe, successful
process, other than to point out (what the inventor does not wish
to conceal) a considerable loss in silver from volatilization.

In the other pyritic methods we have evidence of greater or smaller lead losses, according to the conditions prevailing. Under some circumstances the conditions are not incompatible with the saving of a considerable proportion of that metal—perhaps even a satisfactory amount. We would have generally to consider this question along with such attendant circumstances as the cost of transportation, etc., in order to determine the advisability of recovering all the lead, a part of it, or none at all. As a rule, however, the presence of significant amounts of lead, which we do not wish to sacrifice, is a bar to the employment of the pyritic processes. I regret the inability to furnish statistics showing the lead losses in pyritic work, and can only say that in those charges (comparatively low in lead) which I have worked, the absolute loss seemed to reach from one-third to three-fourths of the whole, and this during the use of the comparatively mild pyritic agencies employed at Mineral.

LOSSES IN SMELTING.

The public are doubtless sufficiently familiar with the principal sources of loss in lead smelting; the same causes are in operation in matting, producing quite similar results. These causes are usually classified as 1, Loss in slags; 2, Loss in flue dust; 3, Loss by volatilization. It is worth while to examine rather closely into the comparative magnitudes of these different losses.

69. SLAG LOSSES.—In the treatment of gold and silver ores the slags carry away on the average as much in the one process as in the other. It is difficult, to be sure, to ascertain exactly what is lost in any given case; the more so, as those conversant with the facts are disinclined to make them public. Without overlooking the natural tendency on the part of smelting people to conceal or

---

NOTE.—1. Without an explicit statement to the contrary, it might be supposed that the pyritic processes are less well adapted to the extraction of copper than of gold and silver, to which they have thus far been principally directed. But in fact, as far as experience shows, it is extracted quite as advantageously. It is probable that the pyritic furnace is able to produce purer mattes than other blast-furnace methods, on account of its greater effects in eliminating arsenic and antimony.

2. The principal advantage of the pyritic treatment of sulphides may be summarized as: First, getting rid of roasting apparatus, and the cost and trouble of running it. Second, getting heat out of the sulphides. Third, getting flux out of them. Fourth, getting matte out of them, which brings down the valuable metals of the charge.

at least minimize their slag losses—a tendency which forewarns us not to attach too much credence to stories of abnormally clean slags—we should endeavor to do the fullest credit to the remarkable performances of the skilled and progressive metallurgists of the day in their efforts to reduce these sources of loss to the lowest practicable point. Of absolutely clean slags there are none. Even the carefully compounded mixtures of the assayer do not yield slags which are absolutely devoid of valuable metals. Assays therefore are never absolutely correct—a fact which may not be known to the assayer, but which the working metallurgist, accustomed to find his silver frequently, and his gold occasionally, "overrun" at the clean-up after a campaign, has the best of evidence to prove.* Regarding therefore the fact that slags must inevitably carry off some portion, however small, of that we work for, an important question confronts the theorist, of how clean can slags in any given case be made? And a more important one arises in the mind of the practical metallurgist, of how clean a slag will it pay to make? We are led by these inquiries in two directions, and we have the general question, what can be done economically in slag formation, which pertains generally to smelting; and second, the question as between matting and lead smelting, which makes under the same conditions the cleaner slags. It is difficult to find examples in real practice where slags are made under conditions so similar as to afford favorable examples for comparison. Perhaps the best practical examples to which I can refer are found in the work of the large lead smelters of the Rocky Mountain region, and the one matte smelter at Denver. Confining the comparison to silver losses we have to compare the loss of one and a half ounces to each ton of slag† at the Argo works with a loss of one ounce, speaking roundly, at the other plants. But the proportion of slag produced to ore smelted would not be more than two-thirds as great in the Argo practice as at the other establishments, where the necessities of the work compel the use of a great deal of flux, which is not used at all at Argo. Accordingly, the silver losses per ton of ore would perhaps

---

* For example, a pile of matte, or of native sulphides, carefully weighed, sampled and assayed, will oftentimes show a gross value less than the same material after roasting, notwithstanding the inevitable losses incident to that operation.

† Private communication from the manager.

be reduced to the same figure. Even cleaner than the lead slags of Colorado are those of Freiberg, where by dint of running all their first slags through the furnace the second time they are reduced to three-tenths of an ounce per ton before being discarded. But the conditions which at Freiberg favor this practice do not prevail in Colorado, where all the requirements of smelting, save very large plants and highly skilled metallurgists, are not to be so easily had; and still less at other localities where smelting has to be carried on under conditions which preclude the thoroughness of extraction which characterizes German practice. At Freiberg and at Swansea it may pay to make exceptionally clean slags, one ton of which does not carry away even so much as one dollar's worth of all valuable metals combined. At Denver the best practice may be to produce slags containing two dollars' worth; while in isolated camps where fuel, labor, refractory materials, etc., are very dear, and the smelting mixture perhaps unfavorable, the best metallurgy may favor comparatively foul slags, carrying perhaps as much as five or more dollars' worth of metals. If the Swansea slags are cleaner than those of Butte, and Clausthal's freer from metals than those made in Eureka or Leadville, it does not follow that the processes or the metallurgy of the latter were at fault. It is always pecuniary profit and not the perfection of processes which is the criterion of metallurgical fitness.

70. LOSSES IN LEAD SLAGS.—Aside from the slag assays given in the Table of Work Done, which are too few to afford decisive evidence as to the points at issue, I have taken pains to collect a large number of others, mainly from the statements of smelting superintendents and chemists, whose truthfulness I assume, from the mass of which I make these deductions:

Three large establishments, dealing extensively with silver ores, which may be taken as the type of the best-conducted lead smelters in the United States, make slags which average, speaking without any attempt at entire accuracy, an ounce of silver per ton, with minute amounts of gold and a small quantity of lead, which are not as significant in this examination. I believe that this class of excellently conducted works do more than half of the lead smelting of the United States. The next class, embracing smaller but not necessarily less well-handled plants, working more restrictedly as to mixture, smelting costs, etc., make slags averaging two to two and a half ounces of silver per ton; and a larger number of concerns, running for the most part intermittently, and engaged oftentimes upon private work for local mines, produce still fouler slags, whose average contents it is impossible to conjecture, but which reaches in some cases five or more ounces. These examples of bad work are typical not of the work of to-day, but of ten or twenty years ago, and are introduced into this discussion with a view of showing what the tendency of the modern practice of lead smelting is toward.

Such being the practical results which are being achieved by the lead smelters, we have to continue our examination and embrace such data as will enlighten us on what the producers of matte can do, and what it is to their interest to do.

71. EFFICIENCY OF LEAD AND MATTE COMPARED.—It is the opinion of Mr. Austin, based upon long experience with both forms of reduction, that the slag losses are practically equal under ordinary conditions. Others, conversant with matting, and perhaps over-enthusiastic with its advantages, have been of the opinion that it saved even a higher proportion, at least of the precious metals, basing their belief upon abnormal results got by the assay of certain experimental slags. I would remark, however, that the lead smelters occasionally produce exceptionally, and even wonderfully clean slags. But these rare exemplars of what we can do, but do not care to repeat, must go for naught in the discussion of such a subject as this. Evidently the skill of the metallurgist, acting upon materials more than ordinarily favorable, is alone to be credited with these exceptional results, and we should not do well to class them with the ordinary run of smelting operations.

On the whole, I am convinced that matte (assuming that it be of the proper chemical and physical constitution) will collect the values (silver and gold) as thoroughly as lead will. But we have somewhat more trouble in removing the matte, now charged with the values, from the contact with the slag; and to this is to be ascribed one of the principal losses which matting is found to incur. When our matte and slag approach each other in specific gravity, so that the difference is not enough to admit of the necessarily complete separation, a loss is bound to take place, and this loss, which is a purely mechanical one, will bear some relation to the difference of the specific gravities. There is not the slightest doubt that the process of sulphidation of the valuable metals during the matting fusion is complete and perfect; but the mechanical separation of the so-formed sulphides presents a

point of inferiority to the separation of lead bullion from its slags
by reason of the greater difference of specific gravity in the latter.

72. INFLUENCES OF SLAG COMPOSITION ON LOSSES.—The
influences of an improperly constituted slag or matte may indeed
be exceedingly detrimental to successful work, especially where
ultra-clean slags and high results are necessary; and in entering
upon matting operations we are not by any means entitled to leave
this phase of the matter to chance. The separation of the
matte from slag is influenced mainly by the following considerations:

First, being dependent upon the difference between the specific
gravity of slag and matte, it is facilitated by the increased gravity
of the matte and by the diminished gravity of the slag.

It is to some extent facilitated by the fluidity of the slag and
made more difficult by its viscosity, although not always and under
all conditions to the extent that might be thought, for, given time
enough, the most viscous slags will release the matte globules.
The presence of solid unmelted stony particles in a slag has little
effect in preventing the separation, and some of the cleanest slags
which are commercially made, contain numerous solid particles
which remain unmelted during the whole operation, and it is only
when the minute particles of sulphide are locked up absolutely
within the solid masses that any loss from the non-fusion need
be apprehended; although these masses may exist in such proportion as to make the slag quite thick and viscid from their
presence. Such slags, it is evident, may only be profitably made
in the reverberatory furnace, where their viscid nature and the
time required for the separation of the matte are not incompatible
with the smelting operation.

The practical questions which arise at this juncture are, what
differences of specific gravities between matte and slag are essential
to a thorough separation, and how can we secure such differences.
The subject, which is not altogether a new one, demands much
fuller treatment than I am able to give it now, the most that I
can say being to give, unsupported by exact data, the conclusions
at which I have arrived.

73. PRACTICAL REQUIREMENTS.—I think that a difference of
one is clearly insufficient for even a tolerable separation under
any circumstances. Lead smelters' slags having a gravity of
four or slightly under, fail to separate satisfactorily from mattes
of five, an average weight. Slags of 3.65 separated, as experiment

proved, very fairly from the same matte. There is no absolute difference of specific gravities which we can assume as essential, because the separation is to some extent contingent on liquidity and upon the time allowed for subsidence. Active boiling and abrupt movements of the fluid mass probably also promote coalescence of the globules and subsidence of the matte.* Because I have never had trouble with the separations when the difference of specific gravity reached 1.75, and because it is probably always possible to achieve that difference, I regard it as advisable to work for it.

An inspection of the list of the slag formers, given previously, shows gravities ranging from two and a half to four, thus having a range of one and a half; while the matte formers, also shown in a previous list, range from four to eight or thereabouts. It appears then that in the effort to attain desirable differences of gravity we can usually effect it easier by making the mattes heavier than the slags lighter. At any rate we are able to control to a considerable extent that source of loss of values which arises from the dissemination of matte particles in slag.

The solution of matte in slags, which has also been considered an important source of loss, has been treated by several writers on copper metallurgy, to whom I shall refer the reader with the remark that the subject does not appear to have advanced much in its theoretic treatment since the days of Le Play, nearly half a century ago. My own views, which are the result of much study, are still of too immature a character to find a place in an essay which is professedly of a practical nature.

74. LOSSES FROM VOLATILIZATION.—The question of the volatility of metals, which plays a great part in general metallurgy, is very important in the various departments of matting, and especially so in pyritic smelting. Herein we are confronted by conditions which determine the formation of oxides, anhydrides, and even of salts, and which favor the sublimation of many of them.

---

* Slow and placid movements of the mixture have been deemed favorable to the subsidence of matte particles. But the escape of globules from moving liquids is conditioned on curvilinear movements, and the tendency to escape varies as the square of the velocity. A high angular velocity is the ideal condition, giving rise to high tangential force. This is only compatible with curved interior surfaces of the forehearth. The ideal form of the basin would therefore approach the spherical, as most conducive to separation, and also the form which the slowly chilling slag tends to produce.

The Bartlett zinc-lead process, which is an extreme form of pyritic smelting, is founded upon the tendency of zinc and lead to volatilize from the furnace, whereby there is formed a sublimate of mixed zinc oxide and lead sulphate, while of the non-volatile substances of the charge, the copper and gold remain as constituents of the matte. We are told that in its practical operation this process causes a loss of from six to fifteen per cent. of the silver, for this metal is also volatile under the conditions which are essential to the sublimation of zinc and lead. These conditions, as far as we know them, are high temperature and the presence of gaseous currents containing oxygen. In the roasting of silver ores, especially in the salt-roasting for chloridation, heavy losses of the same metal are experienced, at times reaching thirty per cent., and in these cases the losses are by some ascribed to the formation of other sublimates, as of arsenic, zinc and antimony, and of chlorine compounds, etc., which act chemically or mechanically to drag away the silver. Silver losses by volatilization from roasting furnaces are experienced at comparatively low temperatures, perhaps not above incipient redness, or even lower; which leads to the question, Is it possible for volatilization to take place at as low a temperature in the upper part of the shaft furnace? If so, then the conditions governing the volatilization may be identical in both processes. In each there is the oxidation of sulphides, arsenides, and antimonides, with the production of oxides of the common metals and the volatilization in a gaseous current of nitrogen and a little oxygen, of sulphurous, arsenious, and antimonious oxides, and likewise of oxides of zinc and lead, should all those elements be present. There is a more rapid gas current in the pyritic shaft, with a shorter exposure of the ore to its influence.

In bessemerizing mattes by the Manhés system, which is an operation chemically identical in principle with pyritic smelting, the conditions are equally favorable for the volatilization of the same substances. Accordingly we might expect to find a corresponding loss of silver, and this in fact is reported to occur in the conversion of argentiferous mattes, although the writer is unable to present well-attested evidence as to the extent of the losses suffered.

Now concerning the silver losses in pyritic smelting, which is an all-important matter in the present connection, I regret being unable to present full information, which is the reader's due, and can only urge in apology the difficulty of securing data from the managers of works, who as a rule are reluctant to make their losses known. Rumor, unverified, makes the loss of silver at one Colorado plant 18 per cent. My own experience in regard to volatilization was exceedingly varied and instructive. Briefly stated, I experienced losses of silver in different campaigns as follows: 15 per cent., 11 per cent., 8.2 per cent., 3.8 per cent.; while in one short campaign the silver recovered "overran" 2.6 per cent. These citations taken by themselves convey no lesson whatever, unless it be that the pyritic process is exceedingly uncertain in its results; but taken in connection with the concurrent circumstances as to charge, blast, and method of work, details upon which I will not dwell, they illustrate a great deal. It is exceedingly difficult in the ordinary run of work to differentiate the losses by slag, by volatilization and by dusting from each other and give each its true measure of responsibility. I may say, however, that at Mineral, where the operations were carried on with particular reference to ascertaining the losses incurred, we were enabled to segregate them in a satisfactory manner, and my conclusions were that about two-thirds of the loss in the first campaign were due to volatilization proper. This great loss was in my opinion the result of experimental and imperfect work in an untried field. The loss of 8.2 per cent. in the third case was about evenly divided between volatilization and slag losses, as shown by regular and frequent slag assays, while in the campaign where 11 per cent. escaped, 3 per cent. appeared to result from sublimation, the rest entering the slag.

The different causes to which I refer the losses were these: First, deficiency of copper in the matte; second, occasional deficiency of the matte, arising from excessive oxidation; third, firetops; fourth, improper composition of the slag; fifth, incomplete separation of matte and slag. Only the second and third of these have a direct bearing on the subject of volatilization losses. Volatilization of silver takes place most copiously when the furnace is run with a hot top; and as the ore at this point has not reached the heat of fusion, and furthermore as chemical action within it has probably not taken place to any extent, at least not to the extent of sulphidation of the precious metals, I conjecture that the silver is lost while yet in the metallic or chloride condition; and I further conjecture that when once in the condition of sulphide it would resist volatilization. I consider that the loss of silver by volatilization depends upon the intensity of the pyritic agencies, and may also be connected with its chemical condition in the ore. It seems probable that silver contained as sulphide intermixed with large quantities of other sulphides in coarse pieces would be less likely to volatilize from the top of the charge than other compounds of a less stable nature merely intermingled mechanically with gangue matter.

It is my impression that neither copper nor gold suffers loss from volatilization while undergoing the pyritic treatment; and in the absence of all testimony upon the matter we may allowably assume from the known characteristics of nickel and cobalt that they also do not. It would appear then that, so far as losses by volatilization are concerned, the pyritic process is better adapted to ores of gold, copper, and probably nickel and cobalt, than to those of silver. And better to silver than to lead.

The last word upon this subject, however, is not yet said. The fact of the volatility of lead affording a ready means of separating that metal from the pyritic charge lets loose a current of speculation as to the possibility of recovering silver also by means of its volatility. It would appear easy to cause the escape of all the silver in the charge by the simple expedient of forming no matte; and could the condensation of the escaping metal be arranged for as cleverly as that of the lead fumes, very possibly a practical process would be established.

I do not find that these volatilization losses differ in cause from those experienced in lead smelting, nor are they greater in degree than were formerly quite common in that pursuit. They arise from ignorance and inexperience, and the very common misconceptio n of furnace effects; and that growth of knowledge and experience which has worked such a transformation in the one art will, I doubt not, do as much for the other. I believe that there is nothing in the ordinary matting processes which renders them intrinsically more liable to volatilization losses than lead smelting, where they are conducted with equal skill and knowledge. As to the pyritic processes I consider that their tendency to volatilize certain elements, while it may, and probably will at first prove a stumbling-block for the smelting practitioner, when the conditions governing the reactions are well understood, so that its results may be foreseen and guarded against, will prove not an unmixed evil. The tendency to volatilize may open entirely new paths in metallurgy, and the lead-zinc process may be the forerunner of a group of methods whose object will be the volatilization and recovery of numerous substances.

75. LOSSES IN FLUE-DUST.—The temporary loss from the formation of flue-dust, which is an unavoidable drawback in lead smelting, is not less so in blast-furnace matting, and the same evil is dealt with in the same way. Spacious flues must always be provided for the recapture of the escaping particles, which are afterward re-smelted with or without the precaution of bricking. There need be no more said upon this point, as there is nothing distinctive in matting practice in this direction.

76. INFLUENCES OF VARIOUS SUBSTANCES UPON EXTRACTION.

Iron.—Increases liquidity and specific gravity of slags. Decreases density of mattes.

Copper.—Increases density of some mattes. Within uncertain limits increases extractive power of mattes for silver and especially for gold.

Lead.—Increases density of slags and mattes. Influence of lead mattes on extraction of the precious metals probably beneficial.

Arsenic.—Increases density, but under some conditions decreases fusibility of mattes. Assists extraction of cobalt and nickel. Is volatilized mainly with formation of arsenious tri-oxide (and loss of silver).

Zinc.—Enters slag as oxide, and matte as sulphide, rendering the former viscid, the latter light, and by injuring the separation diminishes very seriously the saving of the precious metals. By volatilization as metal or oxide causes heavy loss of silver.

Barium.—Enters slag as baryta silicate, and matte (slightly) as sulphide, rendering former heavy though liquid, the latter light, and diminishing the chances of a good separation.

Lime and the Alkalies.—Decrease density of slag, and hence favor separation.

Alumina.—By sometimes rendering slags viscous, prejudices the separation.

Silica.—An excess of silica, by rendering the slag light, favors separation; but by adding to its viscosity injures it. The former quality far outweighs the latter, and some of the cleanest of known slags are extremely siliceous.

77. RÉSUMÉ—1. Lead and properly constituted matte are equally efficient as absorbents of gold and silver, but matte is in certain cases more difficult to separate from the slags.

2. In practical work savings of even 100 per cent. of the assay values are not impossible.

3. The cleanest (of gold and silver) of all known slags are probably those of high silica contents and hence of low specific gravity.

4. Those losses which are caused by the non-separation of matte and slag may be lessened by increasing the difference of the specific gravities of those substances.

5. Slags made by either the lead-smelting or matting methods may carry as little as one dollar per ton in valuable metals (gold, silver, copper and lead).

6. With inferior skill or in the presence of conditions which forbid close work, slags may go as high as five or more dollars per ton.

7. In certain well conducted copper works the slags run but little over one-half per cent. copper; while in others as well conducted they average three times as much.

8. Under appropriate circumstances it may be good metallurgy to sacrifice values in the slag. And while clean slags are an evidence of technical skill, they are not its sole criterion.

9. In smelting practice any important losses of gold by volatilization are unheard of.

10. The volatilization of silver, which may take place in all forms of smelting, is partially under the control of the furnace operator.

11. Losses in unrecovered flue-dust may cover many losses hitherto ascribed to true volatilization.

SALE OF FURNACE PRODUCTS.

78.—Dealings in furnace produce present many peculiarities, a competent knowledge of which is only to be attained by close study and a practical familiarity with the market. This is hardly the place to go into a discussion of the whole subject, however important it may be to the producer of mattes and coppers, but there is a single phase of it which I wish to illustrate, namely, the comparative advantages which the market now offers to the lead smelter and to the matte smelter. To make the points clear I annex a table which contains the elements of several propositions for the purchase of furnace material, on some of which considerable transactions have been carried on by the writer. The matte refiner's charges are of three sorts, which may be systematically grouped as follows:

1. Arbitrary and variable charges upon the ton of matte, or pound of copper; as $10 to $20 per ton of matte, or $2\frac{1}{2}$ to $4\frac{1}{2}$ cents per pound of copper.

2. Discriminations against composition; as a charge per unit of arsenic, antimony, lead, zinc, etc.; or a higher charge for treatment when the value of the precious metals exceeds a certain sum, as $200 to the ton of matte or copper.

3. Standing deductions against the valuable metals in the matte or black copper; as 30 ounces of the silver, 5 to 8 per cent. of the silver, 5 to 13 per cent. of the gold, 1.3 or 1.5 of the copper percentage. Also one-eighth or one-fourth ounce of gold.

The examples to which I apply the bids are two mattes, the one of 50 per cent. copper, with 60 ounces silver and 1 ounce gold —a very common description of product; and the other of 25 copper, with 500 ounces of silver—a decidedly uncommon product, but one with which the writer's practice has made him familiar. For simplicity's sake we will suppose that these substances are devoid of the injurious ingredients, such as arsenic, antimony, or bismuth, upon which additional charges are often based. To parallel the two mattes and afford materials for comparison of the superior advantages of the lead bullion market, I imagine two grades of bullion corresponding to the mattes, one of 60 ounces silver, and 1 ounce gold, the other of 500 ounces silver, no gold; the lead in each to reach 98 per cent. Assuming the market price of copper to be $9\frac{1}{2}$ cents, of lead $3\frac{1}{4}$ cents, of silver 70 cents, we can easily ascertain the aggregate market value which the separated and refined constituents would have, and applying the figures given in the various bids we can arrive at the point to which the table tends, that is, how great a percentage of the total value of the various contained metals would the product bring in the market. The last column on the right shows the percentage of its contents which the more valuable matte and bullion would sell for, the preceding one, the corresponding results for the poorer matte and bullion. It is shown that the selling price at the various markets named, in the case of the 60 ounce, 50 per cent. matte, varies from 50 to 77 per cent. of its total contents; and the very rich silver matte, from 80 to 90 per cent. The bullion brings from 85 to 95 per cent. of the market price of its contained metals. These figures show for one thing that the offer of 0.77 for the common 50 per cent. matte is the highest which has ever been made in this country until of late, for this grade of material, as far as I am aware, although there is reason to believe that very large producers are now able to make better terms for a stated quantity delivered at regular intervals. The richer mattes cannot always be disposed of so advantageously as the table indicates, inasmuch as there is a strong disposition on the part of refiners to discriminate against material very rich in gold and silver. One heavy firm of refiners in Colorado not only make no such discrimination, but add to the producer's convenience by paying cash down for all purchases, which is a vast improvement upon the far Eastern methods.

Much has been and much might still be said as to the defects of the present system (if it can be called a system) of dealings in matte-furnace products; but it is enough at present to declare that we need and must have a better one. The rational and comprehensive methods of the lead bullion buyer may well be our guide in the introduction of improvements; or, in default of such, each matte or copper producer must become his own refiner. With the growth and extension of metallurgical knowledge, and the introducton of simpler and more expeditious processes, adapted to the refining of large or of small outputs, there is more and more reason why furnace produce should be refined and separated at the point of production, and be placed upon the market as finished and purified metals.

TABLE SHOWING PECULIARITIES OF SALE.

Substance on Sale.	Refining Charge — Per Ton.	Refining Charge — Copper, Per Pound.	Deductions on — Silver.	Deductions on — Gold.	Deductions on — Copper.	Deductions on — Lead.	Payment — Copper, Per Unit.	Payment — Lead, Per Pound.	Market.	Quotations — Silver.	Quotations — Copper, Per Pound.	Terms.	Percentage Paid — Poorer Product.	Percentage Paid — Richer Product.
Iron matte	$15		5%	67c. per oz.					Montana.	N. Y. Price.	9⅛c.		71	90
Copper matte		{	30 oz. and 5%	5%	1.8%		$0.95		Montana.	N. Y. Price.	9⅛c. }	Part cash, Balance 60 days	53	84
Copper-silver matte	12		5%	5%	1.5%		0.60		Colorado.	N. Y. Price.	10¾c. }	Cash on delivery.	54	80
Copper matte			7½%	10%	1.3%		1.60 to 1.80		New York.	N. Y. Price.		Three-fourths cash, balance 30 days.	77	90
Copper matte		2⅜c.	7½%	13%	1.3%		N. Y. Price.		New York.	N. Y. Price.		Thirty days.	07	
Copper matte		4½c.	7½%	10%	1.3%		N. Y. Price.		New York.	N. Y. Price.		Thirty days.		
Copper matte	18		7½%	8%	1.5%		0.80			N. Y. Price.		Settlement on buyers' weight, samples and assays.	50	81
Lead bullion	15		1%	67c. per oz.		5%		N. Y. Price.		N. Y. Price.		Cash advanced on bills of lading.	85	95

87

SECTION FOUR.

CHARACTERISTICS OF THE REVERBERATORY MATTING PROCESS.

Applicable Chiefly to:	Special Advantages as Applied to:	Disadvantageous in Case of:
Mixtures containing but small excess of matte-forming substances. Highly siliceous mixtures. Mixtures rich in the alkaline earths. Mixtures rich in alumina. Generally to difficultly fluxed ores, or to those producing viscid or highly refractory slags, and particularly to those containing a significant amount of zinc.	Finely comminuted material. Hot material from calcining furnaces. Mixtures rich in sulphates. Copper ores or furnace produce from which it is desirable to volatilize sulphur, arsenic or antimony.	High cost or poor quality of fire bricks, fire clay and fire sand. Expensive or scarce coal, wood or oil, or other flaming fuel. The presence of important amounts of lead in the ores requiring treatment. Mixtures giving rise to extremely basic or corrosive slags.

CHARACTERISTICS OF THE GERMAN SYSTEM.

Applicable Chiefly to:	Special Advantages as Applied to:	Disadvantageous in Case of:
Mixtures containing no excess of matte-forming ingredients. To very siliceous ores where limestone or dolomite are abundant and cheap.	Mixtures giving rise to extremely basic or corrosive slags. Ores containing a significant proportion of lead, in presence of S, As or Sb.	Costly or scarce coke, charcoal or anthracite. Presence of S, As or Sb, which it is desirable to volatilize.

CHARACTERISTICS OF THE AUSTIN OLDER PROCESS.

Applicable Chiefly to:	Special Advantages as Applied to:	Disadvantageous in Case of:
Mixtures containing a great proportion of matte-forming substances. Siliceous ores capable of forming bi-silicates, mainly of iron. Mixtures containing sulphides in lump form.	Elimination of large quantities of sulphur, arsenic and antimony. Substances tending to form crusts upon the furnace walls.	Ores containing lead which it is desirable to recover.

84 — MATTE SMELTING.

MATTE SMELTING.

CHARACTERISTICS OF THE GRADUAL REDUCTION METHODS—COLD BLAST. (a)

Applicable Chiefly to:	Special Advantages as Applied to:	Disadvantageous in Case of:
Mixtures of moderate tenor in sulphides, etc., and rich in the heavy metals. Mixtures containing copper, with or without gold and silver. Mixtures containing lead, a part of which it is desirable to save. Mixtures giving rise to easily fusible slags.	The use of wood as smelting fuel. The volatilization of a large proportion of the contained sulphur, arsenic and antimony. The concentration of molten matte. (See paragraph 59).	Inability to handle refractory slags. Comparative slowness of the process.

CHARACTERISTICS OF THE GRADUAL REDUCTION METHODS—HOT BLAST. (b)
(HYPOTHETICAL.)

Applicable Chiefly to:	Special Advantages as Applied to:	Disadvantageous in Case of:
Mixtures of higher tenor in sulphides, etc. Mixtures containing any or all of the valuable metals, excepting lead. Mixtures, siliceous and otherwise, of ordinary or exceptional degrees of fusibility.	The probably economical use of wood and coal as smelting fuels. The volatilization of a larger proportion of sulphur, arsenic and antimony.	Additional cost of apparatus

CHARACTERISTICS OF THE BARTLETT PROCESS.

Applicable Chiefly to:	Special Advantages as Applied to:	Disadvantageous in Case of:
Mixtures containing lead and much zinc, little copper, and with or without gold and silver.	Blende ores containing over 20 per cent. zinc. Saving of lead.	Expensive fuel. Heavy loss of silver.

TABLE OF FURNACE EFFECTS—PART I.

Material.	Pyritic Smelting.	Reverberatory Matting.	German System, and Lead Smelting.	[Remarks.
Iron Compounds: (Oxidized) Hematite, limonite, magnetite, roasted sulphides and arsenides.	Reduction by sulphur or carbon to FeO, and slagging as silicates. No metallic iron reduced.	Reduced by sulphur to protoxide and slagged by SiO_2.	Reduced by S or C to protoxide, slagged by SiO_2 or further reduced by C to metallic form, and taken up by S to form matte, or remains as sows.	
Iron: Metallic.	Oxidation to FeO, with evolution of heat, and combination with SiO_2.		Unites with S to form matte. Precipitates metallic lead from its fused sulphides.	
Manganese: Oxides and carbonate.	Reduction to lower oxides and union with SiO_2.	As in pyritic systems.	May enter matte to slight extent. Remainder reduced and slagged.	May replace iron in slags, but not in mattes. (?) Slags very fusible and liquid. Prevents scorification of Zn (Hoffman). Reduced proportion of mattes formed (Iles).
Quartz: Free, or as natural silicates.	In cold blast, must be fluxed mainly by FeO or MnO. In hot blast, processes, may probably be fluxed more largely by alkaline earths, or largely by alumina.	May be fluxed by alkaline earths or alumina, with iron or manganese oxides. May constitute 25 to 70 per cent. of the slag, or even more.	In the German system, may be fluxed as in the reverberatory matting, the slag consisting mainly of lime (and magnesia) silicates, or of more fusible silicates. In lead smelting, must be fluxed with FeO (MnO) and CaO, with silica not below 28 nor above 40 per cent.	Indispensable in all smelting processes. Replaceable to some extent by Al_2O_3, and possibly titanic acid, and by fluorine. For action of latter, see Percy on Copper.
CaO: Calcite and dolomite.	In silicates formed with cold blast CaO must not exceed 27 per cent. of the slag. With hot blast probably more.		Forms silicates. A small portion, reduced to CaS, enters matte. CaO, necessary in Pb-smelting, should not exceed 27 nor fall below 10 per cent.	Said to have a cleansing effect on silver-bearing Pb slags.

TABLE OF FURNACE EFFECTS—PART II.

Material.	Pyritic Smelting.	Reverberatory Matting.	German System and Lead Smelting.	Remarks.
MAGNESIA COMPOUNDS: Dolomite, talc, steatite, etc.	Probably prejudicial to cold blast, but may be handled by hot blast. Effects not studied.	Works slowly, requiring high heats.	Deemed objectionable. High temperatures needed, with high fuel consumption.	High fluxing power, which may offset high fuel consumption.
HEAVY SPAR: $BaSO_4$.	With acid slags and high heats, decomposed with formation of silicate. With basic slags and low heats, eliminated unchanged in part.	Decomposed, with volatilization of SO_3 and slagging of BaO.	Decomposed. Partially reduced to BaS, which enters the matte, partially scorified as silicate. Double decomposition between BaS and $FeSiO_3$, producing FeS and $BaSiO_3$.	Barium slags are of high sp. g. and high liquidity; barium mattes of low sp. g., making separation difficult. In other respects they are advantageously employed in pyritic processes. See Kerl for use of $BaSO_4$ in matting of ores devoid of sulphides.
GYPSUM: $CaSO_4$.	Decomposed. Lime silicate formed. SO_3 volatilized.	As $BaSO_4$.		
LEAD: Oxidized compounds, "carbonates," litharge, etc.	Pb only partially recovered.	Lead slagged. Recovery slight.	Recovered completely as sulphide (German system). Recovered as metallic lead (Lead smelting).	Increases fusibility and increases sp. g. of slags which it enters.
COPPER: Oxidized compounds.	Complete recovery as matte.	Complete recovery as matte.	Complete recovery as matte (German system). Incomplete recovery as metallic copper, with contamination of lead bullion (Lead smelting).	

TABLE OF FURNACE EFFECTS—PART III.

Material.	Pyritic Smelting.	Reverberatory Matting.	German System and Pb Smelting.	Remarks.
COPPER AS SULPHIDE: Chalcopyrite, bornite, fahlore, glance, etc.	Complete recovery as sulphide in matte. S partly volatilized.	As in pyritic processes.	Complete recovery as sulphide in matte. Little or no S volatilized.	
ZINCBLENDE:	Complete decomposition. Zn sublimed and recovered as ZnO in fumes (Bartlett process). Partly sublimed, partly slagged as ZnO, and the remainder driven into matte (Gradual reduction processes).	Mainly driven into matte.	Partially oxidized and slagged, partially matted. Remainder volatilized as Zn, afterward forming ZnO.	Forms troublesome incrustations in blast furnaces.
ZINC: Oxidized compounds.	Behaves as blende in presence of S.	Slagged.	Mainly slagged. Volatilization of a part, as above.	
LEAD SULPHIDE: Galena, jamesonite, lead matte, etc.	Oxidized. Lead largely slagged. Pb volatilized as sulphate and recovered in fumes (Bartlett process).	Oxidation and slagging of lead. Recovery slight.	Complete recovery as sulphide in matte. Decomposed by metallic iron, producing metallic lead (Lead smelting—Precipitation process).	

TABLE OF FURNACE EFFECTS—PART IV.

Material	Pyritic Smelting.	Reverberatory Matting.	German System and Pb Smelting.	Remarks.
ALUMINA:	Difficult of fusion. Injurious to cold-blast process. Handled with more ease by hot-blast methods.	Diminishes fusibility.	For aluminous lead slags, see Keyes in Eighth Report, California Mining Bureau, p. 806.	
PYRITE: FeS_2.	Austin process; One atom sulphur volatilized unoxidized. Iron oxidized with evolution of heat; resulting FeO slagged by SiO_2. Gradual reduction process: Sulphur oxidized to SO_2 and partly to SO_3. Iron oxidized to FeO and slagged. Much heat generated by oxidation. A residuum of undecomposed FeS or other sulphide should remain as matte.	Volatilization of one equivalent of S. Melting of remainder (FeS) as matte.	Absorption of another equivalent of metal to form two molecules of matte (FeS, MS). Occasional volatilization of S as SO_2.	For behavior of roasted pyrite, see "Iron compounds, oxidized."
PYRRHOTITE: Fe_7S_8.	Probably as pyrite.	Probable slight volatilization of S. Remainder melts down as matte.	No volatilization of S. Completely transformed into matte.	As above. See tabulated statistics, "Sudbury."

TABLE OF FURNACE EFFECTS—PART V.

Material.	Pyritic Smelting.	Reverberatory Matting.	German System and Pb Smelting.	Remarks.
ARSENIC: Mispickel, arsenates, arsenites, etc.	Complete decomposition. Oxidation and development of heat. Possible loss of silver by volatilization with As. As probably may be recovered as As_2O_3, in fume, as at Freiberg.	Partial volatilization as sulphide and lower oxide. Incorporation of remainder in matte.	Production of arsenide matte. Contamination of products.	Easily expelled by pyritic processes.
ANTIMONY: Antimonial silver, copper and lead ores.	Behaves as As. Probable volatilization of arsenic and antimony as sulphides in Austin process. Behavior not sufficiently studied.	Behaves as As.	Behaves as As.	Injurious contamination of lead and copper products.
SULPHUR AS SULPHIDES.	Volatilized as elemental S by Austin process. Volatilized as SO_2 and SO_3, by gradual reduction processes, with generation of much heat. Complete elimination of S practicable.	Volatilization of a portion as SO_2 (and SO_3?). Incorporation of remainder in matte. Reactions between oxides (and sulphates) and sulphides eliminate S.	Slight volatilization. Remainder absorbs metals and enters matte.	

TABLE OF WORK DONE.—PART

Localities.	Metal Sought. (Important Metal in Italics.) Assay of Ore Mixture, Per Ton of 2000 lbs.	Character of Ores Treated.
REVERBERATORY MATTING:		
Swansea................	*Copper*	Miscellaneous custom ores, largely calcined fine material............
Black Hawk, 1878........	*Gold, silver,* copper, 3%.....	Miscellaneous custom ores, part heap roasted....................
Argo, Colo., 1892.........	*Gold,* ⅛ to 1 oz.; *silver,* 40 oz.; copper, 3%..........	Miscellaneous custom ores, one-half being crushed and calcined........
Mount Dudley, 1874......	*Silver,* gold, copper.......	Barytic and calcareous ores, low in S and devoid of Fe.............
Butte, 1882..............	*Silver,* gold, copper.......	Siliceous silver ores, high in Zn and Mn compounds..................
Pretoria, S. A. (*a*)........	*Silver,* 30 oz.; copper, 3%...	Basic ferruginous silver ore, containing antimonial oxides and devoid of S......................
Pretoria, S. A. (*b*)........	*Silver,* 30 oz.; *copper,* 3%...	Basic ferruginous silver ore, containing antimonial oxides and devoid of S......................
Butte, 1893..............	*Copper,* 20%; silver, 13 oz...	Roasted copper concentrates, hot from the automatic calciners.....
GERMAN SYSTEM:		
Lend, 1872..............	*Gold,* silver, lead.........	Amalgamation slimes, quartzose lump ore, compact pyrites, roasted matte....................
Mount Dudley, 1874......	*Silver,* gold, copper, lead..	Barytic and calcareous silver ores, low in S and devoid of Fe..........
Butte...................	*Copper,* silver............	Quartz, with much zinc, and little iron and manganese oxides.......
Leadville, 1891..........	*Silver,* copper............	Rich lead slags, 90 parts; copper sulphides, 10 parts..................
Butte, 1893..............	*Copper,* silver............	Copper sulphides, with converter slag, etc.....................
Altai, 186–..............	*Silver, gold*.........	Barytic and siliceous ores..........
Deadwood, 1892.........	*Gold,* silver..............	Auriferous quartz, carrying pyrite..
Sudbury, 1893...........	*Nickel,* copper............	Heap-roasted pyrrhotite and chalcopyrite, diorite gangue..........
Toston, 1887 (*a*).........	*Gold, silver*..............	Siliceous lump ore, with pyrite......
PYRITIC SYSTEMS:		
Leadville, 1892..........	*Silver,* copper............	Cupriferous and argentiferous pyrite....................
Mineral, 1892............	*Silver,* 30 oz.; copper, 1%...	Ferric and manganic oxides and sulphides, in felspathic gangue.......
Mineral, 1893............	*Silver,* 38 oz.; copper, 1%...	Ferric and manganic oxides and sulphides, in felspathic gangue.......
LEAD SMELTING: (*aaa*)		
Leadville, 1880..........	*Silver, lead*...............	"Carbonates," lump form..........
Eureka	*Silver, lead*...............	"Carbonates," lump form, neutral or somewhat basic................
Golden, 1890............	*Gold, silver, lead*.........	Miscellaneous custom ores..........
Pueblo, 1891............	*Silver, gold, lead*.........	Miscellaneous custom ores..........
Tacoma, 1893............	*Gold, silver, lead*.........	Miscellaneous custom ores, largely fine...........................
Freiberg, 1886...........	*Silver, gold, lead*.........	Miscellaneous custom ores..........
Clausthal, 1890..........	....	....

(*a*) Before the inception of pyritic smelting. (*aa*) Cold blast was used at Mineral; the coke, as shown by experiment. (*aaa*) Examples of lead smelting introduced to afford

Î—THE MATERIALS TREATED.

Fluxes Used, with Proportion to 100 Parts Ore	Fuel Used, and Parts to 100 Parts of Charge.
None..	Coal, 45 parts.
Fluorspar, 3 parts...........................	Wood, 1 cord per ton ore.
None..	Coal, 33 parts.
Pyrites, 18 parts; lime, — parts..............	Wood, 1 cord per ton ore.
None..	Wood.
River sand, 25 parts.........................	Coal.
Siliceous and bituminous vein-stuff..........	Coal.
None..	Coal, 30 parts.
Iron ore and lime, 6 parts...................	Charcoal, 25 parts.
Roasted pyrites, etc., 40 parts..............	Charcoal, 35 parts.
None..	None.
None..	Coke, 10.9 parts.
None..	Coke, 10 parts.
Limestone, 13 to 16 parts....................	Charcoal, 70 to 90 parts.
Magnesian limestone, — parts................	Coke.
None..	Coke, 15 parts.
None..	Coke.
None..	Coal, 4 to 6 parts; coke — parts.
None..	Coke and wood. 7½ parts. (aa.)
Limestone, 15 parts..........................	Coke and wood, 9.4 parts.
Iron ore, 8 parts; dolomite, 12 parts.........	Coke and charcoal, about 24 parts.
None..	Charcoal, 25 to 30 parts.
Iron ore, limestone..........................	Coke, 12 to 14 parts.
Iron ore, limestone..........................	Coke and coal, 16.3 parts.
Iron ore, limestone..........................	Coke and charcoal, 14 parts.
Iron ore, 18 parts......,....................	Coke, 8½ parts.
None..	

fuel percentage was calculated on the assumption that 2½ lbs. wood (dry fir) equal 1 lb.
additional grounds of comparison with matting processes.

Localities.	Product.	Weight of Product to Weight of Charge.	Analyses of Products.										P…
			S.	Sb.	As.	Fe.	Cu.	Co.	Ni.	Pb.	Zn.	Ba.	
Reverberatory Matting:													
Swansea	Cu matte	1/5	23	....	....	....	33	....	....	....	....	....	
Black Hawk, 1878	Cu matte	1/10	....	....	....	....	30	....	....	....	....	....	
Argo, Colo., 1892	Cu matte	1/12	....	....	....	....	40	....	....	....	....	....	
Mount Dudley, 1874	Cu matte	1/7.5	....	....	....	....	30	....	....	....	....	....	
Butte, 1882	Cu matte	1/20–1/25	....	....	....	....	53	....	....	....	....	....	
Pretoria, S. A. (*a*)	Cu Sb matte	....	....	....	....	....	....	....	....	....	....	....	..
Pretoria, S. A. (*b*)	Cu Sb matte	1/12	2	38	2	3.6	52	....	....	0.25	....	....	
Butte, 1893	Cu matte	1/5	22	....	....	19	57	....	....	....	....	....	
German System:													
Lend, 1872	Fe matte	3/7	27.9	....	....	55	4.3	....	....	2.1	37*	....	
Mount Dudley, 1874	Cu Pb matte	1/3.5	....	....	....	....	....	....	....	....	....	....	..
Butte	Cu matte	1/5	....	....	....	....	60	....	....	....	....	....	
Leadville, 1891	Cu matte	....	....	....	....	....	20	....	....	....	....	....	
Butte, 1893	Cu matte	1/5	21.5	....	....	20	57	....	....	....	....	....	
Altai, 186–	Fe matte	....	....	....	....	25–35	7–15	....	....	3–8	....	15–20	..
Deadwood, 1892	Fe matte	1/20	....	....	....	....	....	....	....	....	....	....	..
Sudbury, 1893	Cu Ni matte	....	20–30	....	....	25–35	20–25	....	18–23	....	....	....	
Toston, 1887	Fe matte	....	....	....	....	....	....	....	....	....	....	....	..
Pyritic Systems:													
Leadville, 1892	Cu matte	....	....	....	....	14	....	....	....	....	....	....	..
Mineral, 1892	Cu matte	1/12	26	....	....	42	15	....	....	....	4.6	....	
Mineral, 1893	Cu matte	1/15	27	....	....	50	15	....	....	....	....	....	
Lead Smelting:													
Leadville, 1880	Pb bars	1/4–1/5	0.048	0.035	0.22	....	0.07	....	....	98.5	....	....	
Eureka	Fe As matte	1/5.1	3.34	0.13	33	57	1.06	....	....	2.18	.07	....	..
Golden, 1890	Pb bars	....	....	....	....	....	....	....	....	....	....	....	
Pueblo, 1891	Pb bars	....	....	....	....	....	....	....	....	....	....	....	..
Tacoma, 1893	Pb bars	1/5	....	....	....	....	....	....	....	....	....	....	..
Freiberg, 1886	Pb bars	....	....	....	....	....	....	....	....	....	....	....	..
Clausthal, 1890	Pb bars	....	....	0.72	0.06	0.06	0.18	....	0.12	98.8	0.128	....	

* 5⅔Cu_2S, Fe_4…

Assay of Product		Analyses of Slag												Assay of Slag		Authority
Oz. Au.	Oz. Ag.	SiO_2	FeO	MnO	CaO	MgO	BaO	Al_2O_3	ZnO	PbO	Cu_2O	Cu	S	Oz. Au.	Oz. Ag.	
		60.5	28.5		2			2.9				0.55				Le Play.
20-30	600-1000														7	Eggleston.
6	400	41	28	7	7	0.7		3	9	0.5		0.39			1.5	Pearce.
3	1100														7.5	Peters.
2	800	64 (?)														Peters.
		30	70												2	Jennings.
	464	27	55.5		5.32	2.80		4.5			0.45				1.5	Bettel.
0.16	38	33	57					7.8				0.66			0.5	
5	29	51	19.75		15.40	8.57		2.16								Church.
		25	15		18		34								2.6	Peters.
	40															
	93														0.8	Keller.
0.20	36	32	58					8				0.75			0.5	
		50-55	10-18		7-20		5-15	2-4								Kerl.
$250																Browne.
§																
	360	34	48	8	4			3							1.5	Private notes
	560	40.5	23.50	20	4.5			5	3						1.7	Private notes
.26	231									4.5-2.5					2-4	Guyard.
.43	8	26.12	52.80		12			5.8		2.79						Raymond.
																Clark.
																Dwight.
-16	100-200	34													1	Clark.
	150	27	40					7		1.5-4	{0.2 0.3}				0.3	Biddle.
	42	31	40.6	0.86	4.39		0.51	7	8	1.7		0.21	3.55		0.6	Hoffman.

‡ position.

INDEX.

Modern
Copper Smelting.

BY

EDWARD DYER PETERS, Jr.

**Eighth Edition.   Rewritten and Greatly Enlarged.
The standard authority of the world on
Copper Smelting.**

It contains a record of practical experience, with directions how to
build furnaces and how to overcome the various metallurgical difficulties met with in copper smelting.

TABLE OF CONTENTS.

PROFUSELY ILLUSTRATED.   PRICE, $5.00.

THE SCIENTIFIC PUBLISHING COMPANY,

NEW YORK: LONDON:
253 Broadway.   20 Bucklersbury.

www.ingramcontent.com/pod-product-compliance
Lightning Source LLC
Chambersburg PA
CBHW021944190326
41519CB00009B/1135